OXFORD MEDICAL PUBLI

International Classification of Primary Care

ICPC-2

International Classification of Primary Care

Second Edition

Prepared by the Classification Committee of the World Organization of National Colleges, Academies and Academic Associations of General Practitioners/Family Physicians (WONCA), known more briefly as the World Organization of Family Doctors.

OXFORD
UNIVERSITY PRESS

OXFORD
UNIVERSITY PRESS

Great Clarendon Street, Oxford OX2 6DP

Oxford University Press is a department of the University of Oxford.
It furthers the University's objective of excellence in research, scholarship,
and education by publishing worldwide in

Oxford New York

Athens Auckland Bangkok Bogotá Buenos Aires Calcutta
Cape Town Chennai Dar es Salaam Delhi Florence Hong Kong Istanbul
Karachi Kuala Lumpur Madrid Melbourne Mexico City Mumbai
Nairobi Paris São Paulo Singapore Taipei Tokyo Toronto Warsaw

with associated companies in Berlin Ibadan

Oxford is a registered trade mark of Oxford University Press
in the UK and in certain other countries

Published in the United States
by Oxford University Press Inc., New York

First published 1998
Reprinted 1999

A catalogue record for this book is available from the British Library

Library of Congress Cataloging in Publication Data

ICPC-2: international classification of primary care/prepared by
the Classification Committee of the World Organization of National
Colleges, Academies, and Academic Associations of General
Practitioners/Family Physicians (WONCA), known more briefly as the
World Organization of Family Doctors.—2nd ed.
—(Oxford medical publications)
Rev. ed. of: ICPC. 1987.
Cross-referenced to ICD-10
Includes bibliographical references and index.
1. Nosology. 2. Primary care (Medicine)—Classification.
I. ICPC. II. International statistical classification of diseases
and related health problems. III. Series.
[DNLM: 1. Primary Health Care—classification. 2. Episode of
Care. 3. Severity of illness Index. 4. Family Practice-
-classification. WB 15 117 1998]
RB115.136 1998 616.07'076—dc21 97-46472

ISBN 0 19 262802 X

Printed in Great Britain by
Biddles Ltd, Guildford and King's Lynn

Contents

The WONCA classification committee

C. BRIDGES-WEBB, Chairman, Sydney, Australia
B. BENTSEN, Oslo, Norway
N. BENTZEN, Odense, Denmark
R. BERNSTEIN, Ottawa, Canada
N. BOOTH, Newcastle, UK
S. BRAGE, Oslo, Norway
H. BRITT, Sydney, Australia
L. CULPEPPER, Boston, USA
G. FISCHER, Hannover, Germany
T. GARDNER, Dunedin, New Zealand
J. GERVAS, Madrid, Spain
A. GRIMSMO, Surnadal, Norway
J. HUMBERT, Beauvoir sur Mer, France
M. JAMOULLE, Jumet, Belgium
M. KLINKMAN, Michigan, USA
M. KVIST, Turku, Finland
H. LAMBERTS, Amsterdam, The Netherlands
A. LEE, Shatin, Hong Kong
M. LIUKKO, Helsinki, Finland
I. MARSHALL, Mallorca, Spain
F. MENNERAT, Paris, France
L. MICHENER, Durham, USA
G. MILLER, Sydney, Australia
M. MIRZA, Lahore, Pakistan
S. MOHAN, Vijayawada, India
J. NUNES, Mem Martins, Portugal
I. OKKES (formerly Hofmans-Okkes), Amsterdam, The Netherlands
G. PARKERSON, Durham, USA
W. PATTERSON, Edinburgh, Scotland
M. RAJAKUMAR, Kuala Lumpur, Malaysia
D. SALTMAN, Sydney, Australia
P. SIVE, Herzlia, Israel
M. WOOD, Roseland, USA
T. YAMADA, Gifu, Japan
G. ZORZ, Ljubljana, Slovenia

The Chairman,
WONCA Classification Committee
c/o WONCA Office,
Locked Bag 11
Collins Street East PO
Melbourne Victoria 8003
Australia

Fax: +61 3 9650 0236
E-mail: wonca@onaustralia.com.au

Foreword

WONCA has been a Non-governmental Organization (NGO) in official relations with WHO since 1980. Collaboration between the two organizations actually goes back even further, to the first edition of the International Classification of Health Problems in Primary Care (ICHPPC) published by WONCA and the American Hospital Association in 1975 which was based on the Eighth Revision of the International Classification of Diseases (ICD-8). ICHPPC was subsequently updated to be compatible with ICD-9 as ICHPPC-2, published by WONCA and WHO in 1979. ICHPPC-2 was formally endorsed by WHO because of its almost complete correspondence with ICD-9. ICHPPC-2 was republished in 1983, with glossary definitions for most of the terms in the classification, as ICHPPC-2-Defined.

In November 1978 WHO invited a small group of experts experienced in working with classifications of primary health care problems, a majority of whom were members of the WONCA Classification Committee, to participate in a Working Party to develop a classification of reasons why patients make contact with primary health care services, a Reasons for Contact Classification. The group was left free to come up with any plan, concept or suggestion that fitted the purposes, unhampered by the provisions of the Ninth Revision of the International Classification of Diseases (ICD-9). In April 1982 WHO informed the Working Party that the necessary characteristics of the classification were that it should be alphanumeric, have no more than three basic characters in the code, be compatible with ICD-9 and accommodate the disease component of ICD-10 then under development. By 1983, however, it had become clear that ICD-10 probably would not have a format or structure which would correspond with that of the classification being developed by the Working Party, the name of which was amended to International Classification of Primary Care (ICPC).

In November 1985 WHO convened a consultation on Primary Care Classifications. The draft ICPC was reviewed and discussed in the framework of the information needs of WHO Member States for national health development based on the Primary Health Care approach, and WHO activities in the field of health classifications, as well as activities in the area of general medicine at the country level. The meeting concluded that the proposed classification did not reflect the concept, elements or components of primary health care as described in the Alma-Ata declaration and endorsed by the World Health Assembly. Participants felt that the classification could not replace the ICD but could supplement it. As WHO was not willing to publish the classification as it then existed it was published in 1987 by WONCA.

Since that time, ICPC has been increasingly used by individual family physicians and group practices in recording their work, and even by some countries to classify primary care encounters and patients in receipt of national insurance benefit while unable to work due to illness or injury. The recording of both the patient's perceived reason for encounter with a health care provider and the clinician's eventual diagnosis has enabled the data to be used to provide a greater insight into why patients consult their general practitioners and how they present their problems. ICPC has also enabled classification of episodes of illness, and thus broadened the scope of recording the process of general practice/family medicine.

The decision to include a mapping from ICPC-2 to ICD-10 and *vice versa* is warmly welcome even though only about a quarter of the ICPC rubrics can be converted to a single ICD-10 three- or four-character rubric. It is hoped that these conversion codes will prove useful to epidemiologists and other researchers who wish to compare data classified according to ICPC-2 with those from different sources that have used ICD-10. The Division of Health Situation and Trend Assessment, which has responsibility for the development, maintenance and periodic revision of WHO's international health-related classifications, is pleased to have been able to contribute to the ICPC-2/ICD-10 code equivalence lists and to have had the opportunity to provide comments on the final draft of ICPC-2 prior to publication. We look forward to a continuation of the close cooperation and collaboration that has existed between WHO and the WONCA Classification Committee in recent years.

H. R. Hapsara, M.D.
Director
Division of Health Situation and Trend Assessment
World Health Organization

1 Introduction

Historical background

Until the mid 1970s most morbidity data collected in primary care research were classified using the International Classification of Diseases (ICD).[1,2] This had the important advantage of international recognition, aiding comparability of data from different countries. However, there was the disadvantage that the many symptoms and non-disease conditions that present in primary care were difficult to code with this classification, originally designed for application to mortality statistics and with a disease-based structure.

The following early references deal with some of the issues, developments and ideas about general practice classifications in the period before and after the formation of the WONCA Classification Committee in 1972:

- Research Committee of Royal College of General Practitioners. A classification of disease. *J. Roy. Coll. Gen. Pract.* 1959; **2**: 140–59.
- Westbury R C, Tarrant M. Classification of disease in general practice: a comparative study. *Can. Med. Assoc. J.* 1969; **101**: 82–7.
- Bentsen B G. *Illness and general practice: a survey of medical care in an inland population in South-East Norway*. Oslo: Scandinavian University Books, University Press 1970; second edition 1986.
- Hutchinson I M. The Australian morbidity survey 1969–70. *Annals of General Practice* 1971; **16**: 68–72.
- Anderson J E, Leese R E M. Patient morbidity and some patterns of family practice in South-Eastern Ontario. *Can. Med. Assoc. J.* 1975; **113**: 123–6.
- Kjaer P, Mabeck C E, Olsen O M, Pederson P. Testing WONCA's classification of diseases for use in general practice (in Danish). *Ugeskrift Laeger* 1977; **139**: 1614–6.

Recognizing the problems of the ICD, and the need for an internationally recognized classification for general practice, the WONCA Classification Committee designed the International Classification of Health Problems in Primary Care (ICHPPC), first published in 1975,[3] with a second edition in 1979[4] related to the ninth revision of ICD. Although this provided a section for the classification of some undiagnosed symptoms, it was still based on the ICD structure and was still inadequate. A third edition in 1983 had added to it criteria for the use of most of the rubrics[5] greatly adding to the reliability with which it could be used, but not overcoming its deficiencies for primary care. A new classification was needed for both the patient's reason for encounter and the provider's record of the patients problems.

At the 1978 World Health Organization (WHO) Conference on Primary Health Care in Alma Ata,[6] adequate primary health care was recognized as the key to the goal of 'health for all by the year 2000'. Subsequently both WHO and WONCA recognized that the building of appropriate primary care systems to allow the assessment and implementation of health care priorities was only possible if the right information was available to health care planners. This led to the development of new classification systems.

Later in 1978 WHO appointed what became the WHO Working Party for Development of an International Classification of Reasons for Encounter in Primary Care.[7] This group, a majority of whose members were also members of the WONCA Classification Committee, developed a Reason for Encounter Classification (RFEC)[7–9] which later became ICPC.

Reasons for encounter (RFEs) are the agreed statement of the reason(s) why a patient enters the health care system, representing the demand for care by that person. They may be symptoms or complaints (headache or fear of cancer), known diseases (flu or diabetes), requests for preventive or diagnostic services (a blood pressure check or an ECG), a request for treatment (repeat prescription), to get test results, or administrative (a medical certificate). These reasons are usually related to one or more underlying problems which the doctor formulates at the end of the encounter as the conditions that have been treated, which may or may not be the same as the reason for the encounter.

Disease classifications are designed to allow the health care providers' interpretation of a patient's health care problem to be coded in the form of an illness, disease, or injury. In contrast, a Reason for Encounter Classification focuses on data elements from the patient's perspective.[7,10,11] In this respect, it is patient-oriented rather than disease- or provider-oriented. The reason for encounter, or demand for care, given by the patient has to be clarified by the physician or other health worker before there is an attempt to interpret and assess the patient's health problem in terms of a diagnosis, or to make any decision about the process of management and care.

The working group developing the RFE classification tested its several versions in field trials. The first field trial to test the completeness and reliability of the RFEC was a pilot study carried out in The Netherlands in 1980.[8] The results obtained from this pilot study prompted further feasibility testing in 1983. This was carried out in nine countries, namely Australia, Brazil, Barbados, Hungary, Malaysia, The Netherlands, Norway, The Phillipines, and The United States.[9,12,13] The entire classification was translated from English into several languages, including French, Hungarian, Norwegian, Portuguese, and Russian. The analysis of more than 90 000 reasons for encounter recorded during over 75 000 individual encounters and the collective experience of the participants resulted in the development of a more comprehensive classification.[9,12,13]

In the course of this feasibility testing it was noted that the RFEC could easily be used to classify simultaneously the reasons for encounter and two other elements of problem-oriented care, namely the process of care and the health problems diagnosed. Thus this conceptual framework allowed the evolution of the Reason for Encounter Classification into the International Classification of Primary Care.

Problems in relation to the concurrent development of ICD-10 prevented WHO from publishing the RFEC. However, WONCA was able to develop ICPC from it and publish the first edition in 1987. While ICPC-1 was much more appropriate for primary care than previous classifications based on the ICD framework, it did not include inclusion criteria for the rubrics, or any cross-referencing. It was thus in this respect less useful than the previous publication, ICHPPC-2-defined, though it referred to it as a source of inclusion criteria which could be used.

In 1985 a project began in a number of European countries to use the new classification system to produce morbidity data from general practice for national health information systems. This involved translations of the classification and comparative studies across countries. The results were published in 1993 in a book including an update of ICPC.[14]

In 1980 WONCA became a Non-Government Organization (NGO) in official relations with WHO, and joint work together since has led to a better understanding of the requirements of primary care for its own information systems and classifications within an overall framework encompassing all health services.

The international classification of primary care

The International Classification of Primary Care (ICPC[†])[15] broke new ground in the world of classification when it was published in 1987 by WONCA, the World Organization of National Colleges, Academies, and Academic Associations of General Practitioners/Family Physicians, now known more briefly as the World Organization of Family Doctors. For the first time health care providers could classify, using a single classification, three important elements of the health care encounter; reasons for encounter (RFE), diagnoses or problems, and process of care. Linkage of elements permits categorization from the beginning of the encounter with RFE to its conclusion.

The new classification departed from the traditional International Classification of Disease (ICD) chapter format where the axes of its several chapters vary, from body systems (Chapters III, IV, V, VI, VII, VIII, IX, X, XI, XIII and XIV) to aetiology (Chapters I, II, XVII, XIX, XX), and to others (Chapters XV, XVI, XVIII, XXI). This mixture of axes creates confusion, since diagnostic entities can with equal logic be classified in more than one chapter, for example influenza in either the infections chapter or the respiratory chapter, or both. Instead of conforming to this format, the ICPC chapters are all based on body systems, following the principle that localization has precedence over aetiology. The components that are part of each chapter permit considerable specificity for all three elements of the encounter, yet their symmetrical structure and frequently uniform numbering across all chapters facilitate usage even in manual recording systems. The rational and comprehensive structure of ICPC is a compelling reason to consider the classification a model for future international classifications.

Since publication, ICPC has gradually received increasing world recognition as an appropriate classification for general/family practice and primary care, and has been used extensively in some parts of the world, notably in Europe[14] and Australia.[16]

More recently the WONCA Classification Committee has participated in the international development of further initiatives related to classification, including functional status measures, severity of illness indicators, and an international glossary for general/family practice. Information about these is included in this book.

[†]ICPC was first published in 1987.[14] This is now referred to as ICPC-1. In 1993 it was included in a publication about its use in Europe.[13] This is referred to as ICPC-E. This 1998 publication is referred to as ICPC-2. ICPC is used when referring to the generic classification.

Classification, nomenclature, and thesaurus

Labelling aspects of general/family practice, such as reasons for encounter and health problems, requires that the available labels reflect the characteristics of the domain: general practice/family medicine. Labels should be derived from a nomenclature or thesaurus. A nomenclature contains all the terms and professional jargon of medicine, and a thesaurus is a storehouse of terms like an encyclopedia or computer tape with a large index and synonyms.[17]

Classification systems provide a structure to order named objects in classes according to established criteria. They do not necessarily contain all terms, and difficulties arise when they are used as a nomenclature and terms are not found within them. Often many terms are included within one rubric, so that the use of coding based on a classification does not provide adequate specificity.[17]

ICPC is a classification which reflects the characteristic distribution and content of aspects of primary care. It is not a nomenclature. The richness of medicine at the level of the individual patient needs a nomenclature and thesaurus much more extensive than ICPC, particularly for recording the specific detail required in an individual patient record. The use of ICPC together with ICD-10 and other classification systems, such as the Anatomical–Therapeutic–Chemical classification of medications (ATC), can provide the basis of an adequate nomenclature and thesaurus, but if full coding is required these must be supplemented by even more specific coding systems. However, unless such coding systems are based upon a suitable classification, such as ICPC is for general/family practice, it is not possible to extract coherent data about populations rather than just individuals.[17]

Over the years there have been frictions in the relation between the available primary care classifications (ICHPPC and ICPC) and ICD because of conceptual and taxonomical problems. ICD-10, however, now provides a widely recognized nomenclature of diseases and health problems suitable for primary care. Although ICD-10 is not the most appropriate tool for a primary care classification,[18] its use with ICPC as the ordering principle opens a route to good computer-based patient records allowing for the exchange of patient data with other specialists and hospitals.[17]

ICPC-2

This second edition of ICPC has been prepared for two main reasons; to relate it to the tenth edition of ICD, ICD-10, published by WHO in 1992,[2] and to add inclusion criteria and cross-referencing for many of the rubrics. The latter are explained in Chapter 6 and detailed in the tabular list in Chapter 10. In the interests of stability and consistency very few changes to the classification have been made, though many have been suggested, and will be the subject of ongoing work by the WONCA Classification Committee. Feedback from users to assist this process is specially requested.

At the same time this second edition includes information about new developments in the conceptual basis of understanding general/family practice which have arisen in large part from the use of a classification appropriate to the discipline. These are outlined in Chapters 2–5. The book is based on the use of standard terminology as defined in the international glossary published by the WONCA Classification Committee in 1995.[19]

The book also includes information about a number of new initiatives related to classification. The Duke/WONCA Severity of Illness Checklist (DUSOI/WONCA) enables either individual health problems, or the combined health problems of the patient, to be graded in terms of severity (Chapter 7). The COOP/WONCA functional status assessment charts allow assessment of functional status of the patient independent of any particular reason for encounter or health problem (Chapter 8).

The alphabetical index to the tabular list (Chapter 12) is limited to terms from the rubric titles and their inclusion terms. It is not meant to be fully comprehensive (see Chapter 2).

ICPC and ICD

ICPC has always been linked with the widely recognized and used International Classification of Diseases published by the World Health Organization. The first edition contained a list of conversion codes to ICD-9. Since then ICD-10 has been introduced, and ICPC-2 has been carefully mapped to ICD-10 so that conversion systems can be used (Chapter 11). Users who still require a conversion to ICD-9 may obtain a disc from the WONCA Classification Committee. Extensive empirical research has confirmed that ICPC and ICD are complementary rather than in competition.

Translations

WONCA is an international organization and wishes to promote versions of ICPC in languages other than English, which is the working language of the Classification Committee. ICPC has already been translated into 19 languages, and has been published as a book in some of these (Table 1).[13,20,21] There are already several translations

Table 1. Availability of ICPC in different languages[17]

Basque	Hungarian
Danish+	Italian
Dutch+	Japanese+
English+	Norwegian+
Finnish+	Polish
French+	Portuguese+
German	Russian
Greek+	South African
Hebrew	Spanish+
	Swedish

+A separate edition exists in these languages.

of ICPC-2 being undertaken. The committee encourages anyone wishing to promote, assist with, or undertake translations of ICPC-2 to contact them to arrange cooperative work.

The WONCA policy on ICPC-2 translations is:

1. WONCA encourages versions in languages other than English.
2. These must include the whole book, not just the rubrics.
3. There must be no changes to the rubrics. Any extensions must be clearly indicated as such, and approved by the WONCA Classification Committee prior to publication.
4. Translations must be prepared by named translators working in cooperation with the WONCA Classification Committee and to the standards that it sets, particularly in relation to the extent of back translation for checking which may be required.
5. While WONCA will retain the copyright it will usually grant without fee the rights to translating organizations to retain royalties on their versions. This will require a formal agreement between WONCA and the organization or publisher concerned.

Policy on copyright and licensing

The copyright of ICPC, both in hard copy and in electronic form, is owned by WONCA. This policy relates to the electronic version and has the following aims.

Aims

1. To allow the WONCA Classification Committee to promote, distribute, and support ICPC-2, and further develop it as the best classification for primary care.
2. To maintain international comparability of versions of ICPC-2.
3. To obtain feedback and maintain a clearing-house of international experiences with ICPC-2.
4. To achieve recognition of WONCA's initiative and expertise in classification.
5. To promote understanding of appropriate links between ICPC-2 and other classification and coding systems, particularly ICD-10.
6. To encourage use of ICPC-2 rather than inhibit it with restrictions.
7. To obtain financial support to enable achievement of these aims and allow the work of the WONCA Classification Committee to continue and expand.

Policy

1. The electronic version of ICPC-2 should be made available in as many countries as possible.
2. Versions involving additions, translations, or alterations should be made with input from and agreement of the WONCA Classification Committee if they are to be regarded as official WONCA versions.
3. WONCA should license appropriate organizations to promote and distribute electronic versions of ICPC-2 in countries, regions, and language groups.
4. Licence fees may be charged through these organizations to the end users and collected by the distributors for WONCA. The fees will be set by negotiation and may

be waived when there are advantages to WONCA by so doing, such as when use is for research or development.

Readers wishing to obtain this book in electronic form, or incorporate electronic versions of ICPC in computer systems, or develop and use ICPC in other ways should contact a local member of the WONCA Classification Committee (see p. vii) or WONCA (see below).

Abbreviations

As far as possible abbreviations have not been used in this book. In a few places abbreviations which are more commonly used in English than the full expansion, and which are clear in the context, are included in the rubrics. However, some have been needed, and these are as follows:

abn	abnormal
dis	disease
complt	complaint
excl	excludes
incl	includes
NOS	not otherwise specified
sympt	symptom
/	or

User feedback

In order to continue to develop ICPC the WONCA Classification Committee would like to have feedback from as many users as possible with suggestions for clarification, alterations, or extensions. Please contact a local member of the committee (see p. vii) or

The Chairman,
WONCA Classification Committee,
Locked Bag 11,
Collins Street East PO,
Melbourne Victoria 8003,
Australia

Fax: 61 3 9650 0236
e-mail: wonca@onaustralia.com.au

2 The structure of ICPC

ICPC is based on a simple bi-axial structure: 17 chapters based on body systems on one axis, each with an alpha code, and seven identical components with rubrics bearing a two-digit numeric code as the second axis (Fig. 1 and Table 2).

ICPC has a significant mnemonic quality which facilitates its day-to-day use by physicians, and simplifies the centralized manual coding of data recorded elsewhere.

It is presented as a tabular list (Chapter 10). The rubrics for components 1 and 7 are given in full for each chapter. The rubrics in components 2 to 6 are uniform across all chapters and are set out only once. Each rubric has a three-digit code number, a title of limited length, and the codes of the corresponding ICD-10 rubrics. Most rubrics also have inclusion terms, exclusion terms, and 'consider' terms (see Chapter 6). Abbreviations have been used sparingly, and are listed on p. 6. When the word 'multiple' is used in ICPC this refers to three or more.

The alphabetical index to the tabular list (Chapter 12) contains terms from all the rubric titles and their inclusion terms. It is not meant to be fully comprehensive; the terms included are only those which are common or important in primary care. Users seeking terms not included could use the ICD-10 index to find the ICD-10 rubric, and

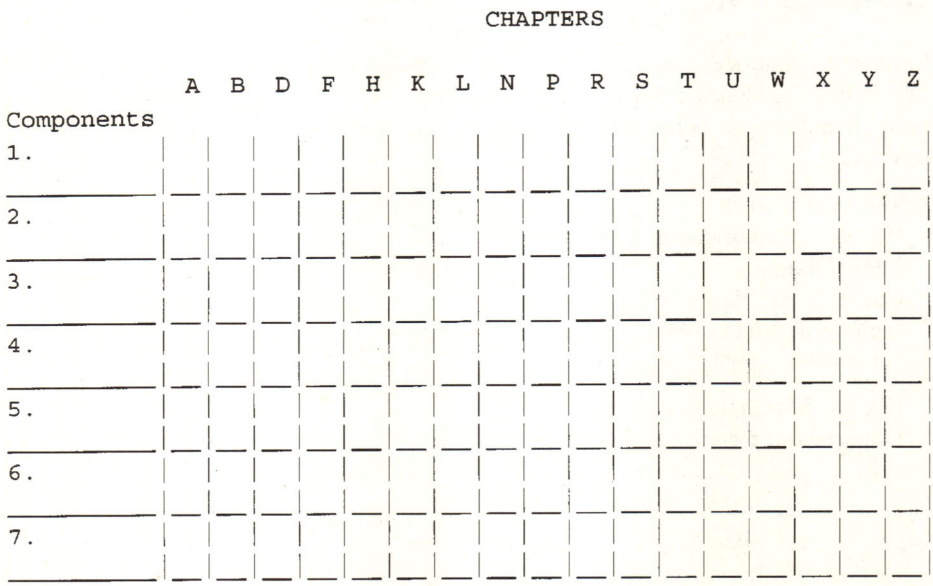

Fig. 1. The structure of ICPC: 17 chapters and 7 components.

Table 2. ICPC chapters and components

A	General and unspecified
B	Blood, blood-forming organs and immune mechanism (spleen, bone marrow)
D	Digestive
F	Eye
H	Ear (Hearing)
K	Circulatory
L	Musculoskeletal (Locomotion)
N	Neurological
P	Psychological
R	Respiratory
S	Skin
T	Endocrine, metabolic and nutritional
U	Urological
W	Pregnancy, child-bearing, family planning (Women)
X	Female genital (X-chromosome)
Y	Male genital (Y-chromosome)
Z	Social problems
Components (standard for each chapter):	
1	Complaint and symptom component
2	Diagnostic, screening, and preventive component
3	Medication, treatment, procedures component
4	Test results component
5	Administrative component
6	Referrals and other reasons for encounter
7	Diagnosis/disease component: —infectious diseases —neoplasms —injuries —congenital anomalies —other

A mnemonic alpha code has been used where possible.

then the conversion tables (Chapter 11) to find the ICPC rubric. A fuller thesaurus in electronic form has been developed by some users, but an approved international version is yet to be developed.

While ICPC is comprehensive enough to allow classification of the main elements of primary care, it still has some limitations. The rubrics in components 2 to 6 covering the process of care are very broad and non-specific. A classification of medications and drugs was developed for and is described in the report of the European study,[14] but is not yet formally included. ICPC does not include objective findings found during physical examination or investigations. These are all matters for further development.

Residual rubrics

Residual rubrics ('rag-bags') are found at the end of a section or subsection; their description includes the word 'other'. Clearly, not otherwise specified (NOS) is implied for all of the terms in these rubrics. A knowledge of the boundaries of each section or subsection is required for the best use of the classification. If in doubt, consult the alphabetical index.

The practical use of morbidity/diagnostic data

Until recently classifications were mainly used for the collection of data for health statistics and formulation of policy. The advent of computer-based medical records has led to even more widespread use as a means of organizing and storing data gathered during routine clinical encounters. These data are needed both as part of the patient medical record and for extraction for health statistics. The classification and coding requirements for those two purposes differ; patient medical records require as much specific detail as possible, whereas health statistics require data which are systematically aggregated into categories based on their frequency or their importance for policy. ICPC was developed for the latter purpose, and must be modified for coding clinical data in medical records.

Optional hierarchical expansion

Clearly, no single international classification can fulfil every need for every user; inevitably users will sometimes want to separate certain problems contained in a single rubric. This usually requires expanded codes using the principle of optional hierarchy. A great deal of expansion is usually needed for coding clinical data in medical records.

It is recommended that whenever possible such expansions conform to the usage in ICD-10, or that ICD-10 codes are used as expansion codes, so that maximum comparability between data systems is maintained. Even then provision for including patient-specific text is needed for adequate specificity for patient care records.[17]

Severity of illness and functional status

Information about severity of illness and functional status assessment of the patient may be recorded in association with use of ICPC, and means of classifying these are therefore included in this book. The Duke/WONCA Severity of Illness Checklist (DUSOI/WONCA) can be related to ICPC rubrics and may be applied to individual health problems, as well as being summed to indicate the severity of the patient's combined health problems (Chapter 7). The COOP/WONCA functional status assessment charts apply to the patient independent of his/her health problems, and are explained in Chapter 8.

3 Episode of care: a central concept of general/family practice

Changes in the need for and use of classifications in primary care have continued since the publication of ICPC in 1987. The main purpose of the classification was then seen to be its use in gathering data for research and policy formulation. However, its use has now widened as research data and practical experiences with ICPC, as well as the emergence of new concepts in general/family medicine, have resulted in new applications.

The most important new applications of the use of ICPC are in describing the construct of episodes of care and in computer patient records. The two are closely related, and depend upon the use of ICPC as the ordering principle of patient data gathered in general/family practice and primary care.

The WONCA definition of general/family practice refers to 'a physician who provides personal, primary, and continuing comprehensive health care to individuals and families'.[19] This is quite similar to that of primary care in the new Institute of Medicine (IOM) definition: 'Primary care is the provision of integrated accessible health care services by clinicians who are accountable for addressing a large majority of personal health care needs, developing a sustained partnership with patients and practising in the context of family and community'.[22]

Episode of care

These definitions have been made operational by choosing the 'episode of care' as the appropriate unit of assessment. Episodes of care are distinguished from episodes of illness or disease in a population. An episode of care is a health problem or disease from its first presentation to a health care provider until the completion of the last encounter for that same health problem or disease (Fig. 2).[17]

Reasons for encounter, health problems/diagnoses, and process of care/interventions form the core of an episode of care consisting of one or more encounters, including changes in their relations over time ('transitions'). An episode of care, consequently, refers to all care provided for a discrete health problem or disease in a particular patient . The 'large majority of personal health care needs', the 'comprehensiveness', the degree of 'integration', of 'accessibility', and of 'accountability' can be assessed when episodes of care are classified with ICPC in a computer-based patient record.

The use of the concept of episode of care was demonstrated in the European study using ICPC.[14] In this study, characteristic epidemiological and clinical similarities and differences between the various sites were established. Also, the concept of reason for

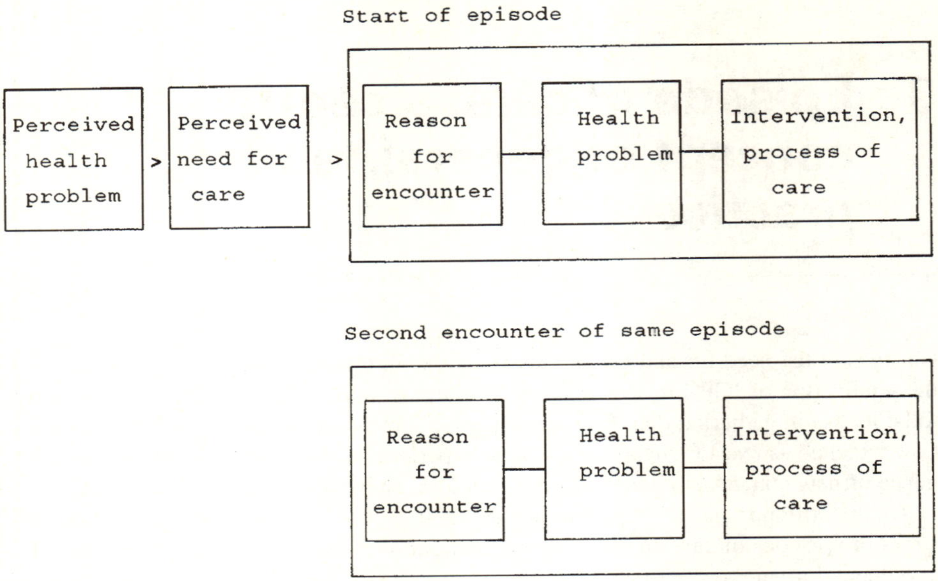

Fig. 2. An episode of care.

encounter proved to be an innovative and practical operationalization of the patient's perspective and demand for care; the validity of the reason for encounter as it was coded by family doctors when compared with the patient's point of view after the encounter, was consistently very high.[23]

The new International Glossary of Primary Care defines the content of general/family practice and gives rules to structure episodes with ICPC in order to allow epidemiological standard retrievals, and to make them comparable in different countries.[19]

Reason for encounter

The reason for encounter (RFE) has been established to be a practical source of patient information, also useful for research and education. This is illustrated by epidemiological data from the Dutch Transition project in the form of standard output, following the rules of the glossary.[17] Beginning with the reason for encounter allows the determination of the probabilities of any given health problem at the start or during follow-up of the episode, per standard sex age group. Thus the top 10 problems related to cough at the start of an episode show clinically important differences between children aged 5–14 and men aged 65–74 (Table 3). The reverse procedure is equally relevant from a clinical point of view: what reasons for encounter were presented at the start and during follow-up of a problem in each standard sex age group? This is illustrated for acute bronchitis (Table 4). These tables document the clinical differences in far more detail than has been possible until now.

Table 3. Top 10 episode titles starting with cough (R05) as the reason for encounter (prior probabilities)

RFE R05 Cough (*N*=1267)	*N*	%
R74 URI (head cold)	456	35.6
R78 Acute bronchitis/bronchiolitis	261	20.4
R05 Cough	159	12.4
R77 Acute laryngitis/tracheitis	110	8.6
A77 Other viral diseases NOS	54	4.2
R96 Asthma	40	3.1
R81 Pneumonia	33	2.6
R75 Sinusitis acute/chronic	30	2.3
R80 Influenza without pneumonia	24	1.9
R71 Whooping cough	22	1.7
Total top 10	1189	92.8
Total	1281	100.0
Men aged 65–74 (*N*=646)	***N***	**%**
R78 Acute bronchitis/bronchiolitis	256	39.1
R74 URI (head cold)	155	23.7
R05 Cough	65	9.9
R77 Acute laryngitis/tracheitis	45	6.9
R75 Sinusitis acute/chronic	22	3.4
K77 Heart failure	15	2.3
R96 Asthma	13	2.0
R91 Chronic bronchitis/bronchiectasis	12	1.8
R81 Pneumonia	10	1.5
R95 Emphysema/COPD	9	1.4
Total top 10	602	92.0
Total	654	100.0

Source: Transition Project, reported in Hofmans-Okkes and Lamberts.[17]

The health problem/diagnosis

The health problem/diagnosis is central to the episode of care and gives it its name. Many health problems are in fact medical diagnoses, but many in primary care are other conditions such as fear of disease, symptoms, complaints, disabilities, or need for care such as immunization. ICPC includes all of these. The health problem may be qualified in terms of its status in the encounter, the certainty which the provider assigns to its diagnosis, and its severity.

The status of the episode in an encounter can be specified as new to both doctor and patient, new to doctor but previously treated outside the current provider system, or neither in the case of follow-up encounters (Fig. 3D). A good computer patient record warns the provider when s/he tries to enter a follow-up encounter for an episode that

Table 4. Top 10 reasons for encounter in an episode of acute bronchitis/bronchiolitis (R78)

Children aged 5–14 (*N*=377)	*N*	%
R05 Cough	321	46.1
A03 Fever	98	14.1
R31 Med exam/health evalua/partial	64	9.2
R02 Shortness of breath/dyspnoea	43	6.2
R74 URI (head cold)	24	3.4
A04 General weakness/tiredness	18	2.6
R03 Wheezing	17	2.4
R64 Provid init episode new/ongoing	17	2.4
R78 Acute bronchitis/bronchiolitis	13	1.9
R21 Sympt/complt throat	9	1.3
Total top 10	624	89.5
Total	697	100.0
Men 65–74 (*N*=422)	*N*	%
R05 Cough	324	39.4
R02 Shortness of breath/dyspnoea	133	16.2
R78 Acute bronchitis/bronchiolitis	100	12.2
R31 Med exam/health evalua/partial	79	9.6
A03 Fever	34	4.1
R25 Abnormal sputum/phlegm	23	2.8
R64 Provid init episode new/ongoing	21	2.6
R74 URI (head cold)	14	1.7
A04 General weakness/tiredness	13	1.6
R01 Pain attrib to respir system	8	1.0
Total top 10	749	91.1
Total	822	100.0

Source: Transition Project, reported in Hofmans-Okkes and Lamberts.[17]

has not yet been established in the database, or whenever a new one is started when an episode with the same title already exists. This is, obviously, vital to ensure the quality of day to day recording.

The extent to which the doctor is certain that his or her diagnosis is correct is another aspect of an episode of care; this can be graded from uncertain to certain, but a standard means of doing this has not yet been adopted. The inclusion criteria for use of rubrics in ICPC-2 will, however, help to ensure that the label chosen for the episode is used consistently by all providers. Pop-up screens can be used to display options at the time of coding in computer-based records.

The third qualification of an episode of care, severity, is discussed in Chapter 7.

Patients with multiple health problems and episodes of care are common in primary care. A good data system will be able to display the interrelationships of these, and provide data on comorbidity (Table 5).

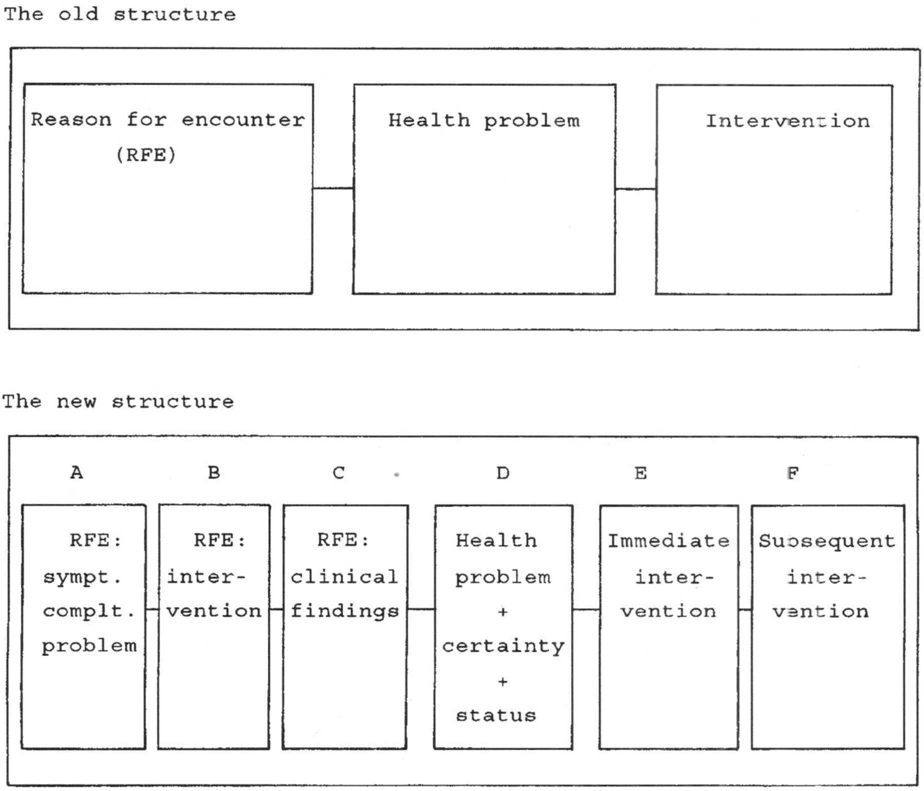

Fig. 3. A new structure for describing encounters.[19]

Interventions, the process of care

The specificity of the three-digit ICPC process code to classify immediate interventions is limited, but usually adequate. However, when drugs are prescribed a drug code is needed. Because of the vast number of medications involved, and the idiosyncracies of national drug availabilities, no internationally suitable code has yet been produced. In Europe an ICPC drug code which is ATC compatible has been valuable and may be suitable for wider adoption.[13]

Patient records

The core of a computer-based patient record is data coded with ICPC which is language independent: this enhances the use of practice records for a comparison of data from different countries, and it supports the development of general/family practice as an internationally well developed profession with a well defined and empirically based frame of reference. The availability of ICPC in 19 languages and the growing number

Table 5. Comorbid episodes for patients with R78, acute bronchitis/bronchiolitis

Children aged 5–14 ($N=329$)	N	%	Prev.
R74 URI (head cold)	90	9.6	274
H71 Acute otitis media/myringitis	57	6.1	173
R78 Acute bronchitis/bronchiolitis	48	5.1	146
R96 Asthma	37	3.9	112
A97 No disease	32	3.4	97
S03 Warts	29	3.1	88
A77 Other viral diseases NOS	21	2.2	64
R76 Tonsillitis acute	20	2.1	61
S18 Laceration/cut	20	2.1	61
D73 Presumed GI infection	17	1.8	52
Total top 10	371	39.6	1128
Total	938	100.0	2851
Mean number of comorbid episodes = 2.9			

Men aged 65–74 ($N=350$)	N	%	Prev.
R78 Acute bronchitis/bronchiolitis	72	4.7	206
A97 No disease	56	3.7	160
R95 Emphysema/COPD	47	3.1	134
K86 Uncomplicated hypertension	46	3.0	131
R74 URI (head cold)	46	3.0	131
K77 Heart failure	35	2.3	100
A85 Adv effect med agent proper dose	30	2.0	86
H81 Excessive ear wax	30	2.0	86
K76 Other/chron ischemic heart dis	30	2.0	86
T90 Diabetes mellitus	25	1.6	71
Total top 10	417	27.4	1191
Total	1521	100.0	4346
Mean number of comorbid episodes = 4.3			

Prev. = Number of comorbid episodes per 1000 patients with R78.
Source: Transition project, reported in Hofmans-Okkes and Lamberts.[17]

of translations of ICD-10 accompanied by alphabetical indexes will allow family doctors in many countries to incorporate a detailed language-specific thesaurus in their system, at the same time using ICPC to systematically structure their records and the database in a more standardized way.

Further developments

The original three basic elements of encounters to be coded with ICPC (reason for encounter, health problem, and interventions) (Fig. 2) have now been expanded into six data entry options (A–F) for computer-based patient records (Fig. 3).[19] The reason for encounter is recorded in two sections: the patient's symptoms and complaints, and the patient's requests for interventions. The clinical findings elicited by the physician in the form of symptoms and complaints are coded in addition to those presented as reasons for encounter. Interventions or processes of care are recorded as immediate (those occurring during the encounter) or subsequent (those which will be done subsequently). Work with these, particularly in The Netherlands, has confirmed the usefulness of the concept of reason for encounter, and further refined the concepts of reason for encounter, health problem/diagnosis, and process of care.[14]

The use of reasons for encounter to estimate prior probabilities is clearly very useful; it can be even more so if reasons for encounter presented by the patient such as cough, shortness of breath, fever, abnormal sputum, or wheezing (Fig. 3A) are distinguished from clinical findings elicited by the physician during history taking (Fig. 3C). ICPC incorporates over 200 symptoms and complaints serving the classification of reasons for encounter and of clinical findings equally well, though it should be noted that it does not yet include a classification of objective findings. Both applications can be included in the encounter and episode structure of a computer-based patient record (Fig. 3A and C). Together they allow a complete calculation of prior probabilities, while the difference between a symptom expressed by the patient as a reason for encounter or elicited by the physician is retained, and the probabilities can be calculated separately if required.

Reasons for encounter in the form of symptoms, complaints or health problems/diagnoses should be distinguished explicitly from those in the form of requests for interventions such as a prescription, an X-ray, a referral, or advice (Fig. 3A and B). Requests for a certain intervention are often followed by this intervention being performed: when patients ask for medication or a blood test, they often receive it.[17] Since patients do actively influence the care provided by general practitioners/family doctors it is important to explicitly document this, also to obtain a better understanding of compliance.

Recording systems should be able to distinguish between diagnostic and therapeutic interventions during the encounter ('immediate', Fig. 3E) and those that will follow ('subsequent', Fig. 3F). The difference between what is in fact being done by the family doctor at the time of the encounter and what is expected to follow is important for the analysis of utilization data, interdoctor variation, and compliance. It also allows better understanding of the shift from prior probabilities in the first encounter of an episode of care to the posterior probabilities during follow-up.[19]

For recording subsequent interventions a more specific process classification than ICPC provides is needed. Development of this is an ongoing activity of the WONCA Classification Committee.

4 Use of ICPC for recording reason for encounter

Procedures for coding information using ICPC vary somewhat according to the type of information being recorded, for example reason for encounter, health problem, or intervention. In order to promote consistent recording and therefore better comparability of data between centres, the following standards are suggested.

Reason for encounter

The primary care provider should identify and clarify the reason for the encounter (RFE) as stated by the patient without making any judgments as to the correctness or accuracy of the reason. This use of the classification is guided by three principles:

1. The reason for encounter should be understood and agreed upon between the patient and the provider and should be recognized by the patient as an acceptable description.
2. The ICPC rubric chosen should be as close as possible to the original statement of the reason given by the patient and must represent a minimal or no transformation by the provider. However, clarification of the patient's reasons for encounter within the framework of ICPC is necessary so that the most appropriate rubric in the classification can be applied.
3. The inclusion criteria listed for rubrics for use in recording health problems/diagnoses are NOT to be used, since the reason for encounter is to be documented from the patient's point of view, based entirely on the patient's statement of the reason.

The way in which a patient expresses his/her reason(s) for encounter determines which chapter and which component to use (Fig. 1 and Table 2). The entire classification is applicable as patients can describe their reasons for seeking health care in the form of symptoms or complaints, as requests for services, or as health problems.

Choosing the chapter code

To code the RFE it is necessary to first select the appropriate organ system or chapter, assign the correct alpha code, and then the two-digit numeric code in the relevant component such as a symptom or complaint, a diagnosis, or an intervention. The alphabetical index should be used when there is uncertainty about the chapter or component in which a specific reason for encounter should be placed. Chapter A is used for reasons for encounter which relate to unspecified or multiple body systems.

When ICPC is used for recording RFE four rules apply to the use of the chapters, and three rules to the use of components. Those rules are listed below with examples of the application of those rules.

Rule 1

The reason for the encounter should be coded as specifically as possible and may require some clarification by the provider.

Example

Chest pain can be coded as A11 (chest pain not otherwise specified (NOS)), or as K01 (pain attributed to heart), or as R01 (pain respiratory), or as L04 (chest symptoms/complaints). The decision as to the correct selection is not based on the opinion of the provider as to the type of chest pain but, rather, to the manner in which the patient expresses his/her reason for encounter when clarification is sought by the provider.

'Its all over my chest...'	A11
'My chest hurts when I cough'	R01
'I have chest pain... I think its my heart'	K01
'I have chest pain after falling down stairs'	L04

Rule 2

Whenever the patient makes a specific statement use his/her terminology.

Example

Jaundice, in the form of a diagnostic descriptive term can be found in Chapter D (digestive) but the patient may present this symptom as a yellow discoloration of the skin (Chapter S). If the patient expresses the problem as 'jaundice', the ICPC code is D13. If, however, the patient states 'my skin has gone yellow' the correct code would be S08, regardless of the fact that the health care provider is positive that the diagnosis is some form of hepatitis.

Rule 3

When the patient is unable to describe his/her complaint, the reason given by the accompanying person is acceptable as that stated by the patient (e.g. a mother bringing in a child or relatives accompanying an unconscious patient).

Rule 4

Any problem whatsoever presented verbally by the patient should be recorded as a reason for encounter. Multiple coding is required if the patient gives more than one reason. Code every reason presented at whatever stage in the encounter it occurs.

Example

'I need my blood pressure tablets. Also my breasts are tender and sore' —K50, X18. If later the patients asks 'What is this lump on my skin?' that is also coded as a reason for encounter —S04.

Choosing the component code

1. Symptoms and complaints

The most common reasons patients give for seeking health care are presented in the form of symptoms and complaints.[14,16,23,24] Therefore, it is expected that Component 1 (symptoms and complaints) will be used extensively. These symptoms are specific for each chapter; nausea is found in the Digestive chapter (D09), while sneezing (R07) is located in the Respiratory chapter. While most of the entries in this component are symptoms specific to the chapter in which they are found, some standardization has been introduced for ease of coding.

Throughout most of the chapters, with the exception of psychological and social, the first rubric(s) relate to the symptom pain. Examples of these are earache (H01) and headache (N01). There are also four standard Component 1 rubrics in each chapter. They are:

- 26 Fear of cancer
- 27 Fear of having a disease or condition
- 28 Limited function/disability
- 29 Other symptoms/complaints

Codes 26 and 27, and sometimes also a few others, are used when the patient expresses concern about or fear of cancer or some other condition or disease. Examples are:

'I'm afraid I have TB'	—A27
'I'm worried that I have cancer of the breast'	—X26
'I'm scared of venereal disease'	—Y25

Even though the provider thinks that such an expressed fear is unwarranted or illogical, it constitutes the patient's reason for encounter.

Rubric —28 should be used when the patient's reason for encounter is expressed in terms of a disability which affects activities of daily life and social functions.

Examples

'I cannot climb stairs because of the cast they have put on my leg for my fractured ankle' —L28 (Component 1) and L76 (Component 7).

'I can't work in the office because I can't sit for any length of time because of my hemorrhoids' —K28 (Component 1) and K96 (Component 7).

In each chapter the component code 29 is the residual or 'rag-bag' rubric for symptoms/complaints. This contains uncommon and unusual symptoms and complaints which do not have a separate rubric, and is also appropriate for symptoms/complaints which are not clearly stated. The index should be checked for synonymous terms in other rubrics before using this rubric.

2. Diagnostic, screening and preventive procedures

The reasons included in this concept are those in which the patient seeks some sort of procedure, such as 'I'm here to have a blood test' (—34). The patient may request a particular procedure in connection with an expressed problem or as a single demand, such as

'I want the doctor to examine my heart' K31, or
'I think I need to have my urine tested'(—35), or
'I've come for the result of my X-ray'(—60), or
'I need a vaccination'(—44).

Clarification by the provider is necessary to find out why the patient thinks he or she needs a urine test in order to select the appropriate alpha code. If it is because of a possible bladder infection the code is U35; if because of diabetes T35. If the result of an X-ray which is being requested refers to a barium meal D60. A request for vaccination against rubella A44.

3. Medication, treatment, procedures

These reasons are expressed when the patient requests a treatment or when the patient refers to the physician's instructions to return for specific treatment, procedure, or medication as the reason for encounter. Further clarification by the provider is often necessary in order to identify the most appropriate code.

Examples

'I need my medication' (—50). If the patient expresses the reason why he is taking the medication or the provider knows the reason, select the appropriate alpha code, e.g. for a sinus infection the code would be R50.

'I'm here to have my cast removed' (—54). If it is evident that, for instance, the patient had a fracture of the left arm the correct alpha code to select would be L.

'I was told to come for removal of the stitches today' (—54). Although at first one might assume that all suture removal would be in the Skin chapter, the patient might have stitches from eyelid surgery F54 or from a phimosis operation Y54.

4. Test results

This component should be used when the patient is specifically requesting the results of tests previously carried out. The fact that the results of the test may be negative does not affect the use of this component. Often the patient will request the test result and its consequences and seek more information on the underlying problem. In that case, also consider using the additional code —45 (health education, advice).

Examples

'I need the results of my blood test'. If the test was for anaemia code B60, if for lipids T60, if the patient cannot specify A60.

'I want to know what they found on the X-rays of my stomach that were taken last week' (D60).

'I am supposed to pick up the result of my urine test and take it to the urologist. I also want to know what he will do and which examinations and treatment I can expect' (U60, U45).

5. Administrative

Administrative reasons for encounter with the health care system include such things as examinations required by a third party (someone other than the patient), insurance forms which require completion, and discussions regarding the transfer of records.

Examples

'I need this medical insurance form completed' (A62).

'My fracture is healed and I need a certificate to go back to work' (L62).

6. Referrals and other reasons for encounter

If the patient's reason for encounter is to be referred to another provider —66, —67, and —68 can be used for this purpose. If the patient states his/her reason for the encounter is 'being told by you to come back', or 'being sent by someone else', use —64 or —65.

When a provider initiates a new episode or takes the initiative for the follow-up of an already existing episode of a health problem such as hypertension, obesity, alcoholism, or a smoking habit, it will be appropriate to code the reason for encounter as —64.

Example

A patient presenting with a blocked ear due to earwax, which is removed, has his blood pressure measured and found to be high, and also receives advice about smoking. The patient's reasons for encounter and the related problems and treatment would be recorded as follows:

H13 (blocked feeling in ear), H81 (earwax), H51 (removal of earwax).

K64 (provider initiated), K85 (raised blood pressure), K31 (checking of blood pressure).

P64 (provider initiated), P17 (tobacco abuse), P45 (advice to stop smoking).

7. Diagnosis/disease

Only when the patient expresses the reason for encounter as a specific diagnosis or disease should it be coded in Component 7. The reason for encounter of a patient who is known to be a diabetic but comes in complaining of weakness should not be coded to diabetes but to the problem expressed: weakness (A04). However, if the patient states that he has come about his diabetes the diagnosis 'diabetes' should be coded as his reason for encounter (T90).

If the patient names a reason for encounter in the form of a diagnosis which the provider knows is not correct, the 'wrong' RFE of the patient is coded rather than the 'correct' one of the physician; for example a patient presenting with a reason for encounter of 'migraine', when the provider knows it is tension headache, or a patient who is known to have nasal polyps presenting with 'hayfever'.

Examples

'I am here because of my hypertension' (K86).

'I come every month for the arthritis of my hip' (L89).

Rules for components

The following rules for the use of each component will reinforce the description of the components.

Rule 1

Whenever a code is shown preceded by a dash (—), select the chapter code (alpha). Use A when no specific chapter can be selected, or when multiple chapters are involved. All codes must begin with an alpha code to be complete.

Example

Biopsy will be coded —52, for digestive system D52, for skin S52. Medication prescribed will be coded as —50. A patient requesting medication for asthma R50.

Rule 2

Rubrics from more than one component, or more than one rubric from the same component, can be used for the same encounter if more than one reason is presented by the patient.

Example

'I've had abdominal pain since last night and I vomited several times' D01, D10.

'I have some abdominal pain and I think that I may have appendicitis' D06, D88.

5 Use of ICPC for recording health problems and process of care (interventions)

Health problems

ICPC can be used to record the provider's assessment of the patient's health problems. This can be done in terms of symptoms and complaints, or diagnoses, so can be derived from Component 1 or Component 7. The latter is based on the lists of diseases, injuries, and related health problems in the International Classification of Diseases (ICD), but includes as separate rubrics only those that are common or important in primary care.

Many of the health problems which are managed in primary care cannot be designated in terms of disease or injury. They include symptoms and complaints, which are listed in Component 1. Sometimes there is no apparent health problem involved in an episode of care, as when it relates to need for immunization or a Pap smear or advice. These episodes can be labelled using rubrics such as A97 No Disease, or A98 Health Maintenance/Preventive Medicine.

In Components 1 and 7 the corresponding ICD-10 codes are listed for each rubric. Sometimes these are an exact one-to-one match, but more often there are several ICD codes for an ICPC-2 rubric, and sometimes there are several ICPC-2 codes for a single ICD-10 rubric. A full conversion structure is given in Chapter 10.

In order to improve reliability of coding health problems using ICPC-2, many of the rubrics in Component 7 have inclusion criteria specified. These are explained in Chapter 6.

Rubrics in Components 1 and 7 often have additional information as a guide to their use: lists of synonyms and alternative descriptions as inclusion terms; lists of similar conditions which should be coded elsewhere as exclusion terms; and lists of less specific codes which might be considered if the particular patient's condition does not meet the inclusion criteria. There are no such guidelines for rubrics in the process Components 2 to 6.

General rules for coding health problems

Users are encouraged to record during each encounter, the full spectrum of problems managed, including organic, psychological, and social health problems, in the form of episode(s) of care. Recording should be at the highest level of diagnostic refinement for which the user can be confident, and which meets the inclusion criteria for that rubric.

In any data system it is necessary to have clear and specific criteria for the way in which health problems or episodes of care are recorded. This applies particularly to the relationship between the underlying condition and manifestations when both may be available as rubrics in the classification, and is best illustrated by an example. A patient with ischaemic heart disease may also have atrial fibrillation and resulting anxiety. It should be policy to include as separate episodes of care manifestations which require different management, and in the above example the atrial fibrillation and anxiety would be recorded as additional episodes of care.

Some systems require that problems be coded only from components 1 and 7; others also accept codes from other components, so that if, for example, the patient attends for a tetanus immunization without a current injury, the problem could be coded as N44.

In ICPC localization within a body system takes precedence over aetiology, so that when coding a condition which because of its aetiology can be found in several chapters (for example, trauma) the appropriate chapter should be used. Chapter A (general) should be considered only if the site is not specified or if the disease affects more than two body systems. All chapters provide specific rubrics based on the body system or organ involved in the disease and the aetiology. Conditions accompanying and affecting pregnancy or the puerperium are usually coded to Chapter W, but a condition is not coded to Chapter W merely because the patient is pregnant; it should be coded to the appropriate rubric in the chapter representing the body system involved. All social problems, whether identified as a reason for encounter or as a problem, are listed in the first component of Chapter Z.

Specific rules for coding health problems using inclusion criteria (see also Chapter 6)

1. Coding of diagnoses should occur at the highest level of specificity possible for that patient encounter.
2. Inclusion criteria contain the minimum number of criteria necessary to permit coding with that rubric.
3. Consult the criteria *after* the diagnosis has been formulated. They are NOT guidelines for diagnosis, NOR are they intended to be used as a guide to therapeutic decisions.
4. If the criteria cannot be fulfilled, consult other less specific rubrics suggested by the term 'consider'.
5. For those rubrics without inclusion criteria, consult the list of inclusion terms in the rubric, and take into account any exclusion terms.

Process of care, interventions

ICPC can be used to classify the interventions used in the process of medical care with Components 2, 3, 5, and part of Component 6; however, Component 4 and some rubrics of Component 6, namely, —63, —64, —65, and —69, cannot be used in this way.

These process rubrics are broad and general, rather than specific. For instance, a blood test (—34), even if relating to only one body system (e.g. cardiovascular, K34), may encompass a great variety of different tests such as enzymes, lipids, or electrolytes.

The process codes in Components 2, 3, and 5 follow the major headings to be found in the far more detailed IC-Process-PC, which was developed by the WONCA Classification Committee.[25] ICPC and IC-Process-PC are, therefore, compatible one with the other. The details found in IC-Process-PC may be applied to the three-digit ICPC codes by expanding to four or five digits.

In Components 2, 3, 5, and the part of Component 6 which can be used to classify the process of care, the rubric codes are standard throughout the chapters at the two-digit level. The alpha code of the correct chapter has to be added by the provider who is doing the coding. A limited number of rubrics in the first and seventh components of Chapters W, X, and Y also contain procedures such as delivery, abortion, and family planning.

The most important principle in the coding process is to code all those interventions which take place during that particular encounter and which have a logical relation to the episode of care. A fourth or fifth digit may be necessary for increased specificity, as in the following examples:

Example 1

—54	Repair/fixation/suture/cast/prosthetic device
L54.1	Application of casts
L54.2	Removal of casts

Example 2

—40	Diagnostic endoscopy
—D40	Diagnostic endoscopy of the digestive system
—D40.1	Gastroscopy

More than one process code may be used for each encounter, but it is extremely important to be consistent. For instance, measuring the blood pressure, which is routine for hypertension, can be coded as K31 on every occasion. Routine examinations, complete or partial, both for body systems or for the general chapter must also be coded with consistency. Below are examples of definitions for complete and partial examinations which have been used in one setting. However, it is essential that each country develops a definition of what constitutes a 'complete examination—general' and a 'complete examination—body system' for that culture and that these definitions are used consistently. This will ensure that what is contained in each 'partial examination—general' or 'partial examination—body system', in that country will also have consistency.

Complete examination

The term 'complete examination' refers to an examination which contains those elements of professional assessment which by consensus of a group of local professionals reflects the usual standard of care. This examination will be complete with regard to either the body system (e.g. eye, Chapter F) or as a complete general examination (Chapter A).

Partial examination

The term 'partial examination' in any chapter refers to a partial examination directed to the appropriate specific organ system or function. When more than two systems are involved in a limited or incomplete examination it is designated general (Chapter A). Most encounters will include a partial examination to evaluate acute and simple illnesses or return visits for chronic illnesses. The following are examples:

Complete examination—general, general check-up = A30
Complete neurological examination = N30
Partial examination—general, limited check on several body systems such as respiratory and cardiovascular = A31
Partial examination—body system, measuring blood pressure = K31

The following procedures are regarded by the WONCA Classification Committee as included in routine examinations to be coded in rubrics —30 and —31 rather than coded separately:

- inspection, palpation, percussion, auscultation
- visual acuity and fundoscopy
- otoscopy
- vibration sense (tuning fork examination)
- vestibular function (excluding calormetric tests)
- digital rectal and vaginal examination
- vaginal speculum examination
- blood pressure recording
- indirect laryngoscopy
- height/weight.

All other examinations are to be included in other rubrics.

Component 2—diagnostic, screening and preventive procedures

Diagnostic and preventive procedures cover a wide range of health care activities including immunizations, screening, risk appraisal, education, and counselling.

Component 3—medications, treatment, procedures

This component is designed to classify those procedures done on site by the primary care provider. It is not intended that it be used to document procedures done by providers to whom the patient has been referred, for which a much more extensive list of procedures would be required. Immunizations are coded in Component 2.

Component 4—test results

Component 4 does not relate to process or interventions.

Component 5—administrative

This component is designed to classify those instances where the provision of a written document or form by the provider for the patient or other agency is warranted by existing regulations, laws, or customs. Writing a referral letter is only considered to be an administrative service when it is the sole activity performed during the encounter, otherwise it is included in Component 6.

Component 6—referrals, and other reasons for encounter

Referrals to other primary care providers, physicians, hospitals, clinics, or agencies for therapeutic or counselling purposes, are to be coded in this component. Referrals for an X-ray or a laboratory investigation should be coded in Component 2.

For more specificity, a fourth digit can be added, for example:

—66 Referral to other provider/nurse/therapist/social worker.
- —66.1 Nurse
- —66.2 Physiotherapist
- —66.3 Social worker

—67 Specialist
- —67.1 Internist
- —67.2 Cardiologist
- —67.3 Surgeon.

6 Inclusion criteria in ICPC

Introduction

It has always been clear to the WONCA Classification Committee that an internationally agreed list of rubrics to classify problems met in primary care would not in itself ensure the highest level of statistical comparability. In the International Classification of Health Problems in Primary Care (ICHPPC-2-Defined) published in 1983 inclusion criteria for the use of each rubric were introduced to improve consistency of coding.[5]

Inclusion criteria are not the same as definitions. They should be considered in relation to their purpose, to improve consistency of coding, rather than as definitions for delineating health problems. We have, however, tried to ensure that they are compatible with accepted definitions, such as those in the International Nomenclature of Diseases (IND).

In this publication many of the inclusion criteria originating in ICHPPC-2-Defined have been updated and are directly related to ICPC rubrics. In some instances, new or extensively modified inclusion criteria have been created based on the theoretical framework described in the next section. Although this publication marks an advance in the taxonomy of general/family practice, it is not yet ideal. ICPC is a classification very much in evolution, and experience with the inclusion criteria presented in this volume will undoubtedly lead to further refinement in the years to come. We welcome comments from users.

Theoretical framework for assignment of inclusion criteria

The theoretical framework used to assign inclusion criteria in this classification is based on the presence of four general categories of diagnosis in primary care: aetiological and pathological disease entities, pathophysiological conditions, nosological diagnoses (syndromes), and symptom diagnoses. It was decided to apply different principles to each category based on its characteristics:

- *Aetiological and pathological*: the diagnosis has proven pathology or aetiology; inclusion criteria are based on standard disease definition, with modification where necessary to allow application to general/family practice.
  Examples: appendicitis, acute myocardial infarction.
- *Pathophysiological*: the diagnosis has a proven pathophysiological substrate; inclusion criteria include symptoms, complaints, and characteristic objective findings.
  Examples: presbyacusis, hypertension.
- *Nosological*: the diagnosis depends on a symptom complex based on consensus between physicians, without a proven pathological or pathophysiological base or

aetiology, and is often called a syndrome; inclusion criteria include only symptoms and complaints.
Examples: depression, irritable bowel syndrome.

- *Symptom*: a symptom or complaint is the best medical label for the episode.
Examples: fatigue, eye pain.

The criteria

The underlying principle used was to provide THE MOST CONCISE INCLUSION CRITERIA POSSIBLE WHICH WOULD MINIMIZE VARIABILITY IN CODING. Adherence to this principle led to the use of *minimal inclusion criteria* for each rubric. This requires further explanation.

For most diagnostic rubrics, the reader will find one or more criteria which must be fulfilled to code a problem under that title. Sometimes there is a choice of criteria; at other times criteria from a list must be met. When 'or' is used in a list it is with its inclusive meaning, which is the same as 'and/or'. 'Multiple' in this book means three or more.

Attempts were made to specify the *minimum criteria* needed in order to reduce the complexity of coding and thus minimize miscoding. In addition, we have only included those criteria which have sufficient *discriminatory value* to distinguish one rubric from another with which it might be confused. In some cases, the available criteria may be too few to exclude *all* other possible conditions which might be coded mistakenly to a particular rubric, but they will exclude the common ones.

The criteria have whenever possible been based on clinical criteria, rather than requiring the results of tests and investigations. They are as far as possible independent of technology, which varies considerably throughout the world, and is rapidly changing. This makes them appropriate for primary care use throughout the world.

This approach is very different from that seen in classic disease-oriented textbooks, which usually list all signs and symptoms, or all potential criteria, associated with a particular diagnostic title. We believe that in order to maximize the utility of criteria-based problem coding in general practice, brevity must supersede exhaustiveness.

Sometimes the rubric title is itself adequately specific. In these cases, no inclusion criteria are given. To avoid errors, each rubric, with inclusion and exclusion terms, and inclusion criteria, should be read in its entirety.

Attempts were not made to provide criteria for every rubric, particularly residual rubrics, which contain too many disparate diagnoses for useful definition. In these cases, the reader should consult the list of diagnoses included in the rubric title and inclusion terms, or refer to the more complete list given for the relevant rubrics in ICD-10.

Cross-referencing

As well as inclusion criteria, each rubric may have the following information:

- includes: a list of synonyms and alternative descriptions which are included in the rubric

- excludes: a list of similar conditions which should be coded elsewhere, with the appropriate code for each
- consider: a list of rubrics with their codes, usually less specific, which might be considered if the particular patient's condition does not meet the inclusion criteria

Advantages of this framework

The use of this framework results in clear and generally accepted inclusion criteria for problems which are common in general/family practice, and which require inclusion criteria if they are to be coded consistently.

Another major advantage of this framework is that employing minimal inclusion criteria results in coding procedures which are easy to learn and apply in the real world of general/family practice. This will reduce the magnitude of the problem of intercoder variation.

Using inclusion criteria

Inclusion criteria SHOULD NOT be used when recording patient reason for encounter, since these should be coded in terms of what the doctor understands the patient to say, irrespective of whether or not the patient is 'correct'.

Inclusion criteria SHOULD be used when coding the diagnoses or problems the doctor manages. Even when the problem has to be coded only as a symptom or complaint, some guidance may be needed in order to select the most appropriate code. For example, feeling faint (N17) is not coded in the same rubric as actual fainting (A05); and abdominal pain may be generalized (D01), epigastric (D02), or localized in other regions (D06). The options need to be clear to users so that the most appropriate alternative is used.

Applying the criteria at different stages of the problem

The inclusion criteria are primarily designed to code the early presentation of a problem. If the problem is to be coded during a later encounter (after its modification by time or therapy) the coder should consider the historical information (e.g. blood pressure may well be normal at later consultations in a patient with hypertension receiving therapy but the condition would still be coded as hypertension).

Disadvantages of the system

Clearly, this system of inclusion criteria is not without hazard. In order to improve the accuracy and reliability of statistics from general/family practice, hard edges have been put to diagnostic concepts, many of which seem, in reality, to have blurred borders. Although sharp borders may not be needed for therapy or management, accurate data are needed for purposes of research. The use of hard-edged inclusion criteria may increase the content of residual less specific rubrics, but this is preferred to making most rubrics non-specific. For coding problems which do not fully meet the given criteria, less specific alternatives are suggested following 'Consider:'. These suggestions are in addition to those items which are listed as exclusions in the rubric.

Some possible misconceptions

It is important that the reader clearly understand several things which the criteria are NOT intended to do.

1. *They do not serve as a guide to diagnosis.* The primary purpose of the classification is to reduce chances of miscoding *after* a diagnosis has been made, and not to eliminate the possibility of diagnostic error. The assumption is that the user will have considered the differential diagnosis prior to the time of coding. In most cases good practice of medicine requires far more information than is given in the inclusion criteria to make accurate diagnoses.
2. *They do not set standards for care.* Although information derived from the use of the classification may change medical concepts and ultimately impact on standards of care, these inclusion criteria are intended solely to improve the quality of data recording.
3. *They do not act as a guide for therapy.* The criteria given for inclusion or exclusion for a condition do not necessarily relate to the criteria for use of various therapies. For example, the practitioner may well decide that therapy for migraine is indicated in a patient whose findings were insufficient to fulfil the criteria listed under that diagnostic title, and whose condition is coded as 'headache'.

Sources

The Committee felt no compulsion to devise new definitions and based inclusion criteria on existing ones, if appropriate for the objectives given above. In fact, few existing definitions did meet those requirements because most had been prepared for research projects rather than for clinical practice and so tended to be rather cumbersome. However, the inclusion criteria included here are compatible with most standard definitions of diseases.

If someone else's work has been used inadvertently without acknowledgement, apologies are given: imitation is the sincerest form of flattery.

7 Severity of illness coding

Development of severity of illness coding

Since 1993 the WONCA Classification Committee has been developing the Duke Severity of Illness Checklist (DUSOI) system[26] for international use. The WONCA Severity of Illness Field Trial (WONCA-SIFT) was conducted to test the system in 16 countries.[27] The committee recognized that a method is needed to enable doctors to code not only the name of each health problem, but also the level of severity of each problem. This is applicable to problems whether in respect of episode of care, or for each encounter (see Fig. 2).

The ICPC is now unique among international classification systems in that it can be used to classify health problems by their level of severity in the individual patient with the health problem. The severity coding system, the Duke/WONCA Severity of Illness Checklist (DUSOI/WONCA), is an extension that enables the physician or other health worker not only to give the problem a standardized title and classification code, but also a standardized severity code that indicates which patients with the same health problem have the more or less severe problem. Since the severity parameters and criteria of the system are generic, not health problem specific, they can be applied to any health problem. This generic quality also allows comparison of the severity of different health problems based upon the same standards for assessment. The system is feasible for use by family/general practitioners in the clinical setting as demonstrated in the WONCA-SIFT field trial.[27]

Coding severity of illness

The DUSOI/WONCA severity of illness coding system allows ICPC to be used to classify health problems in terms of severity. To code severity, the health care provider identifies each problem at the time of patient encounter and determines how severe each problem is for that particular patient at that particular time. Severity is based upon the following four generic parameters:

1. *Symptoms* during the past week.
2. *Complications* during the past week.
3. *Prognosis* during the next six months if no treatment were to be given for the health problem.
4. *Treatability*, or the need for treatment and the expected response to treatment by this patient.

An example of a completed form is shown in Fig. 4, in which the provider, John Smith, has listed the current health problems which he addressed during the encounter

DUKE/WONCA SEVERITY OF ILLNESS CHECKLIST: **DUSOI / WONCA** * Patient: Mary Jones

Birthdate: Nov. 6, 1925 Female: ✓ Male: __ Provider: John Smith Date of Encounter: Oct. 5, 1995

Health Problems (Addressed during this encounter)	Raw Scores (Enter 0-4)** Symptoms	Complications	Prognosis	Treatability	Total Raw Score (0-16)	Severity Code*** (0-4)	ICPC Code
EXAMPLE: Gout	3	1	3	2	9	3	T92:3
1. Ischaemic Heart Disease without Angina	2	0	4	2	8	2	K76:2
2. Diabetes Mellitus	0	0	2	2	4	1	T90:1
3. Acute Bronchitis	3	0	2	2	7	2	R78:2
4.							
5.							
6.							

(Use additional pages if more than six health problems.)

** RAW SCORES

	None	Questionable	Mild	Moderate	Major
1. Symptoms (past week):	0	1	2	3	4
2. Complications (past week):	0	1	2	3	4

	Disability: None	Mild	Moderate	Major	Threat to Life
3. Prognosis (next 6 months, without treatment):	0	1	2	3	4

	Need for Treatment: No	Questionable	IF YES →→	Expected Response to Treatment: Good	Questionable	Poor
4. Treatability:	0	1		2	3	4

*** SEVERITY CODES

Total Raw Score		Severity Code	Severity
0	=	0	None
1 - 4	=	1	Mild
5 - 8	=	2	Intermediate
9 - 12	=	3	Moderate
13 - 16	=	4	Maximum

Fig. 4. Duke/WONCA severity of illness checklist: DUSOI/WONCA.

for patient Mary Jones on 5 October 1995. Gout is printed on all forms as an example; it does not apply to this patient. If patient Jones had had gout, the provider would have listed gout again and scored its severity according to its specific effects on Mary Jones. Actually, at this encounter this patient had ischaemic heart disease without angina, diabetes mellitus, and acute bronchitis.

Raw scores

The raw severity scores that are possible for each of the four severity parameters are shown in the large box at the bottom of the DUSOI/WONCA form. In the example of ischaemic heart disease without angina (K76) in Fig. 4, the severity of *symptoms* was rated '2' on a scale of 0 to 4, because the patient was considered to have mild symptoms at that time. Severity in terms of complications was rated '0' because no complications for ischaemic heart disease were evident clinically. For DUSOI/WONCA ratings, the definition of a complication is 'a health problem which is secondary to another health problem, but which is not listed and rated as a separate problem'. If a *complication* is recorded on the form as a separate health problem, the effects of this separately recorded complication should not be included in the severity rating of the health problem of origin, in order to avoid assigning double weight to a complication in the scoring. *Prognosis* for ischaemic heart disease with angina in Fig. 4 was scored '4' because the provider made a clinical judgment that this patient would be expected to die during the next six months following the encounter if no treatment were given, thereby allowing the heart disease to have full effect on the untreated patient. If Dr Smith had predicted that Mary Jones would not die without treatment, but rather would experience major disability, then a rating of '3' for prognosis would have been appropriate. Disability is defined as 'any limitation of a person's ability to function in everyday life'. Major disability (raw score 3) is defined as 'much restriction of usual activity and much care needed from others'. Mild disability (raw score 1) is defined as 'little restriction of usual activities', and moderate disability (raw score 2) is defined as 'much restriction of usual activity but little care needed from others'. *Treatability* was rated '2' in the example because the provider decided that this particular patient needed treatment and would be expected to have a good response to that treatment.

Severity codes

To determine the single-digit DUSOI/WONCA severity code, the raw scores for each health problem are summed, and the total raw score is converted to a severity code using the conversion table in the small box at the bottom of the form. In the example of ischaemic heart disease without angina in Fig. 4, the total raw score = 8(2+0+4+2), and the severity code = 2. (The conversion table shows that raw scores of 5 to 8 = a severity code of 2.) The severity code of 2 indicates that ischaemic heart disease without angina in this particular patient at this particular encounter is of intermediate severity, on a scale of 0 to 4 from 'none' to 'maximum' severity.

The severity code can be added to the problem code as an extension, using ':' as the link, a convention which distinguishes the severity extension from other extensions which may be used. Hence the code for the ischaemic heart disease without angina in Fig. 4 is K76:2.

Results of the severity of illness field trial (WONCA-SIFT)

The international study was conducted during a two-year period (1993–5) to test the reliability, feasibility, and potential clinical usefulness of the DUSOI/WONCA. Initially 47 general/family practitioners from 16 different countries participated. Of these, 22 practitioners from 9 countries (Belgium, Germany, Hong Kong, Israel, Japan, The Netherlands, Spain, The United Kingdom, and The United States) completed data collection.[27]

The 22 practitioners performed DUSOI/WONCA ratings on 1191 patients. The study group had a mean age of 59.2 years; 59.6% were females; and they had a total of 2488 health problems. Reliability of the DUSOI/WONCA was estimated from ratings on a series of standardized health problems. The intraclass correlation coefficient (ICC)[28] for interrater reliability was 0.45 and the ICC for intrarater reliability ranged from a low of 0.39 for the social problem of partner being ill (ICPC code Z14), to 0.78 for obesity (ICPC code T82) and 0.68 for anxiety (ICPC code P74). Feasibility for use in practice was good, as indicated by an average of only 1.9 minutes required to rate the DUSOI/WONCA on each patient (ranging from less than 1 to 10 minutes). The physicians experienced no difficulty in using the system in 71.1% of patients. They found it quite useful in 14.7% of patients, somewhat useful in 53.6%, and of no use in 31.7%. Usefulness was higher in patients with higher severity of illness scores.

The mean DUSOI/WONCA severity score for all 2488 health problems was 39.1 (scale = 0–100 from lowest to highest severity), and the problems were distributed among the five severity classification codes as follows: Code 0 (no severity) = 1.6%, Code 1 (mild severity) = 29.9%, Code 2 (intermediate severity) = 45.9%, Code 3 (moderate severity) = 19.3%, and Code 4 (maximum severity) = 3.3%. Wide variations in severity were shown, both between different diagnoses and within each diagnosis. For example, mean severity for respiratory health problems varied between 26.4 for upper respiratory infection (URI, ICPC code R74) to 53.2 for chronic obstructive pulmonary disease (COPD, ICPC code R95). For URI, the frequency of severity codes ranged from 61.1% for Code 1 to 0% for Code 4, in contrast to COPD with the range from 8.4% for Code 1 to 10.6% with Code 4.

When surveyed at the end of the field trial concerning their future anticipated personal use of the DUSOI/WONCA, 41.2% of the 22 participants reported they might use it in patient care, 71.2% might use it in research, 43.8% in teaching, and 52.9% in practice management.

It was concluded that the DUSOI/WONCA is feasible and potentially useful clinically in family/general practice.[27] Although the practitioners were not queried about the usefulness of the system for disease severity classification, the empirical findings of the field trial indicate that it is well suited for this purpose.

8 Functional status assessment: the COOP/WONCA charts

In 1987 the WONCA Classification Committee began to develop a way of classifying and recording the overall functional status of the patient as distinct from the status of severity of their health problem(s).[29] Over a number of years this work, later conducted in cooperation with the WONCA Research Committee, resulted in production of the COOP/WONCA Functional Status Assessment Charts.[30,31]

Functional status is a measure of an individual's overall well-being. It is one of the set of global measures of health status, which also include assessments of clinical status and quality of life. The International Glossary for General/Family Practice defines functional status as 'the ability of a person to perform and adapt to his/her environment, measured both objectively and subjectively over a stated period of time'.[19] Implicit in any definition of functional status is the importance of factors other than disease in the health of patients. As the complexity and chronicity of medical problems increase, general/family practitioners will become more reliant on indicators of functioning as well as disease status to monitor their interventions and measure health outcomes.

Functional status relates to the patient, not to the health problem, disease, or episode of care. It thus relates less directly to the ICPC codes than does severity of illness. However, its importance in general/family practice warrants its inclusion in this book.

For some time general practitioners have recognized the integral importance of health promotion and the measurement of functional status in consultations. These measurements are particularly important in dealing with ageing and those with chronic problems. The addition of functional status measures to the recording of reason for encounter, diagnosis, and therapeutic interventions is a logical step for the process of classification in general/family practice.

Instruments for measuring functional status

One of the first instruments to be recognized by WONCA as a reliable and practical measure of functional status in the family practice setting was the Dartmouth COOP Functional Assessment Charts.[32] These charts were modified by the classification committee and promoted for use in conjunction with ICPC. The revised charts are known as the COOP/WONCA charts.

The COOP/WONCA charts, whilst specifically developed for general/family practice, are not the only instruments available for assessing functional status. There are a plethora of indicators currently available. Several have been used in general practice settings.

The Medical Outcomes Trust Short Form 36-item inventory and derivatives of this instrument have been widely used in primary care settings. Similarly, the Duke Health Profile has been used successfully in North American settings.[33] In Europe, several other instruments have been used. The Sickness Impact Profile (SIP) and the Nottingham Health Profile (NHP) are the two most widely cited. Some of these instruments were designed for research not clinical purposes, (e.g. the Sickness Impact Profile[34]).

To date, the COOP/WONCA charts have been tested most extensively in general/family practice settings.[35] Internationally, they have been found to have good face validity and clinical utility in general practice.[36] General practitioners have found the charts easy to use within the consultation and helpful as measures of overall patient status and as outcomes of care.

With any measure of functional status, cultural and context issues need to be explored. Some studies of the charts have suggested that they do not exhibit cross-cultural stability. As a research instrument the test–retest reliability will always be an issue for indicators that are global and influenced by so many variables. Several studies have looked at these issues. Standardization of test conditions and assessment of intrarater reliability may improve the results for research projects.

COOP/WONCA charts

The current form of the COOP/WONCA charts was determined through extensive testing in general/family practice settings. There are now six charts: physical fitness; feelings; daily activities; social activities; change in health; and overall health. An example of the Daily Activities Chart is shown in Fig. 5. Additional charts for pain and sleep are under development.

Each chart consists of a lead sentence with five options for response. Pictorial depictions of the five possible responses accompany the text. These drawings have enhanced the applicability of these charts in settings where there is variability of literacy amongst the general practice patient population.

To date the charts have been published in the following languages: Chinese, Danish, Dutch, Finnish, French, German, Hebrew, Italian, Japanese, Korean, Norwegian, Portuguese, Spanish (Catalan, Castilian, and Callego), Slovak, Swedish, and Urdu.[31]

Use of the charts

The charts can be used independently or in groups. When more than one chart is used it is recommended that they are administered in the following order: physical fitness, feelings, daily activities, social activities, change in health, overall health. The preferred method of use of these charts is self-administration. However, one study has shown a correlation between self-assessment and provider assessment.[31] The average time for completion of the six charts is less than five minutes.

When the charts are used in new cultural settings, it is important to establish that the concepts measured are appropriate and specific to that environment. Appropriate translation is the first step.

Daily activities

During the past 2 weeks...
How much difficulty have you had doing your usual activities or tasks, both inside and outside the house,because of your physical and emotional health?

No difficulty at all		1
A little bit of difficulty		2
Some difficulty		3
Much difficulty		4
Could not do		5

Fig. 5. COOP/WONCA Functional Health Status Chart: daily activities.

Measuring functional health status with the COOP/WONCA Charts: a Manual,[31] provides further information about the development and use of the charts, how to translate the charts, and a contact list for further assistance, including authors of the various translations.

Relationship between ICPC and the COOP/WONCA charts

Together with ICPC the COOP/WONCA charts can be used to explore the relationship between functionality and health problems. For example, Rubric 28 of Component 1 (symptoms and complaints) of all chapters of ICPC refers to limited function and disabilities. Functional status could be coded in this component with the addition of an extra digit. However, since functional status relates to the patient as a whole and not to the health problem, the relationship becomes difficult to interpret when there is more than one active problem, because comorbidity complicates the interpretation. For example, hypertension and diabetes in one patient can both impact on functional status, but their relative importance and effects cannot be determined from routine recording. Even with only one problem, functional status measures go beyond assessing problem status and therefore their relationship to a particular ICPC code may not be straightforward.

9 References

1. International classification of diseases (9th revision). Geneva, World Health Organization, 1977.
2. International Statistical Classification of Diseases and Related Health Problems (10th revision). Geneva, World Health Organization, 1992.
3. International Classification of Health Problems in Primary Care (ICHPPC). Chicago, World Organization National Colleges, Academies and Academic Associations of General Practitioners/Family Physicians (WONCA)/American Hospital Association (AHA), 1975.
4. ICHPPC-2 (International Classification of Health Problems in Primary Care). Oxford, Oxford University Press, 1979.
5. ICHPPC-2-Defined: International Classification of Health Problems in Primary Care, 3rd edition. Oxford, Oxford University Press, 1983.
6. Report of the International Conference on Primary Care, Alma Ata, USSR, 6–12 September 1978; WHO/Alma Ata/78.10.
7. Meads, S. The WHO Reason for Encounter classification. *WHO Chronicle* 1983; **37**(5): 159–162.
8. Lamberts H, Meads S, and Wood M. Classification of reasons why persons seek primary care: pilot study of a new system. *Public Health Rep.* 1984; **99**: 597–605.
9. Lamberts H, Meads S, and Wood M. Results of the international field trial with the Reason for Encounter Classification (RFEC). *Med. Sociale Preventive* 1985; **30**: 80–87.
10. Working party to develop a classification of the 'Reasons for Contact with Primary Health Care Services'. Report to the World Health Organization, Geneva, Switzerland, 1981.
11. Wood M. Family medicine classification systems in evolution. *J. Fam. Pract.* 1981; **12**: 199–200.
12. Lamberts H, Meads S, and Wood M. Results of the field trial with the Reason for Encounter Classification (RFEC). In: Cote R A, Protti A J, and Scherner J R (ed.) Role of Informatics in Health Data Coding and Classification Systems. Amsterdam, Elsevier/JFIP-JMIA, 1985.
13. Bentsen B G. International Classification of Primary Care. *Scand. J. Prim. Health Care* 1986; **4**: 43–56.
14. Lamberts H, Wood M, Hofmans-Okkes I (ed.). The International Classification of Primary Care in the European Community: with Multi-Language Layer. Oxford, Oxford University Press, 1993.
15. Lamberts H and Wood M (ed.). ICPC: International Classification of Primary Care. Oxford, Oxford University Press, 1987.

16. Bridges-Webb C, Britt H, Miles D A, Neary S, Charles J, and Traynor V. Morbidity and treatment in general practice in Australia 1990–1991. *Med. J. Aust.* 1992; **157**, Suppl. 19 Oct: S1–S56.
17. Hofmans-Okkes I M and Lamberts H. The International Classification of Primary Care (ICPC): new applications in research and computer based patient records in family practice. *Fam. Pract.* 1996; **13**: 294–302.
18. Wood M, Lamberts H, Meijer J S, and Hofmans-Okkes I M. The conversion between ICPC and ICD-10: requirements for a family of classification systems in the next decade. In: Lamberts H, Wood M, and Hofman-Okkes I (ed.) The International Classification of Primary Care in the European Community: with Multi-Language Layer. Oxford, Oxford University Press, 1993: 18–24.
19. Bentzen N (ed.). An international glossary for general/family practice. *Fam. Pract.* 1995; **12**: 341–369.
20. Jamoulle M and Roland M. Classification Internationale des Soins Primaires (traduction francaise de l'ICPC). Edition Alexandre Lacassagne, Lyon, 1992.
21. Jamoulle M and Roland M. Approches taxonomiques en medicine de famille, assorties d'une terminologie medicale normalisee et classifiee a usage informatique en soins de sante primaires. CARE Editions, Brussels, 2 vol., 1996.
22. Donaldson M S, Yordy K D, Lohr K N, and Vanselow N A (ed.). Primary care: America's health in a new era. Washington DC, National Academy Press, 1996.
23. Hofmans-Okkes I M. An international study into the concept and validity of the 'reason for encounter'. In: Lamberts H, Wood M, and Hofmans-Okkes I M (ed.) The International Classification of Primary Care in the European Community. Oxford, Oxford University Press, 1993: 34–44.
24. Nylenna, M. Why do our patients see us? A study of reasons for encounter in general practice. *Scand. J. Prim. Health Care* 1985; **3**: 155–162.
25. International Classification of Process in Primary Care (IC-Process-PC). Oxford, Oxford University Press, 1986.
26. Parkerson G R Jr, Broadhead W E, and Tse C-K J. The Duke Severity of Ilness Checklist (DUSOI) for measurement of severity and comorbidity. *J. Clin. Epidemiol.* 1993; **46**: 379–393.
27. Parkerson G R Jr, Bridges-Webb C, Gervas J, Hofmans-Okkes I, Lamberts H, Froom J, Fischer G, Meyboom-de Jong B, Klinkman M, and Maeseneer J. Classification of severity of health problems in family/general practice: an international field trial. *Fam. Pract.* 1996; **13**: 303–309.
28. Shrout P E and Fleiss J L. Intraclass correlations: uses in assessing rater reliability. *Psychol. Bull.* 1979; **86**: 420–428.
29. WONCA Classification Committee. Functional status in primary care. New York, Springer, 1990.
30. Scholten J H G and van Weel C. Functional Status Assessment in Family Practice. MEDITekst CIP-Gegevens Koninklikje Bibliotheek, The Hague, 1992.
31. van Weel C, Konig-Zahn C, Touw-Otten F W M M, van Duijn N P, and Meyboom-de Jong B. Measuring functional health status with the COOP/WONCA Charts: a manual. CIP-Gegevens Koninklijke Bibiliotheek, The Hague, 1995.

32. Nelson E C, Wasson J, Kirk J, *et al.* Assessment of function in routine clinical practice. Description of the COOP chart method and preliminary findings. *J. Chron. Dis.* 1987; **40** (Suppl. 1): 55s–64s.
33. Parkerson G R Jr, Broadhead W E, and Tse C-K J. The Duke Health Profile, a 17-item measure of health and dysfunction. *Med. Care* 1990; **28**: 1056–1072.
34. Bergner M, Bobbitt R A, Carter W B, *et al.* The Sickness Impact Profile. Conceptual formulation and methodology for the development of a health status measure. *Int. J. Health Serv.* 1976; **6**: 393.
35. Hutchinson A, Bentzen N, Konig-Zahn C (eds.) Cross cultural health outcome assessment: a user's guide. *European Research Group on Health Outcomes* (ERGHO), 1997; 1–184.
36. Bentsen B, Natvig B, Winnem M. Assessment of own functional capacity. COOP–WONCA charts in clinical work and research (in Norwegian, English summary). *Tiddsk. Nor. Laegeforen.* 1997; **117**: 1790–93.

10 International Classification of Primary Care: tabular list

The tabular list consists of details of all the rubrics in ICPC. The process components 2–6, which are standard in all chapters, are set out first, followed by components 1 and 7 in which each rubric is specific in each chapter.

Standard process components of ICPC: components 2–6

The dash (—) shown in first position must be replaced with the appropriate alpha code for each chapter.

Component 2—Diagnostic, screening and preventive procedures

—30 Medical examination/health evaluation—complete
—31 Medical examination/health evaluation—partial
—32 Sensitivity test
—33 Microbiological/immunological test
—34 Blood test
—35 Urine test
—36 Faeces test
—37 Histological/exfoliative cytology
—38 Other laboratory test NEC
—39 Physical function test
—40 Diagnostic endoscopy
—41 Diagnostic radiology/imaging
—42 Electrical tracings
—43 Other diagnostic procedures
—44 Preventive immunizations/medications
—45 Observation/health education/advice/diet
—46 Consultation with primary care provider
—47 Consultation with specialist
—48 Clarification/discussion of patient's RFE/demand
—49 Other preventive procedures

Component 3—Medication, treatment, procedures

—50 Medication–prescription/request/renewal/injection
—51 Incision/drainage/flushing/aspiration/removal body fluid (*excl.* catheterization—53)

—52 Excision/removal tissue/biopsy/destruction/debridement/cauterization
—53 Instrumentation/catheterization/intubation/dilation
—54 Repair/fixation–suture/cast/prosthetic device (apply/remove)
—55 Local injection/infiltration
—56 Dressing/pressure/compression/tamponade
—57 Physical medicine/rehabilitation
—58 Therapeutic counselling/listening
—59 Other therapeutic procedures/minor surgery, NEC

Component 4—Test results

—60 Results tests/procedures
—61 Results examination/test/record/letter from other provider

Component 5—Administrative

—62 Administrative procedure

Component 6—Referrals and other reasons for encounter

—63 Follow-up encounter unspecified
—64 Encounter/problem initiated by provider
—65 Encounter/problem initiated by other than patient/provider
—66 Referral to other provider/nurse/therapist/social worker (*excl.* medical)
—67 Referral to physician/specialist/clinic/hospital
—68 Other referrals NEC
—69 Other reason for encounter NEC

Layout of rubrics in components 1 and 7

Rubrics include information under several headings. The ICD-10 codes which are listed for each rubric sometimes relate also to one or more other ICPC-2 rubrics (see Chapter 11). Rubrics are set out in the following format:

Code Title **ICD-10 code(s)**

incl: *terms included*
excl: *terms excluded, with their ICPC codes*
criteria: *criteria for inclusion in this rubric*
consider: *rubrics to be considered if the criteria are not met*

Example:

A73 Malaria ***B50 to B54***

incl: complications of malaria
excl: malaria prophylaxis A44
criteria: intermittent fever with chills and rigors in a resident of, or recent visitor to, a malarial region; or demonstration of malarial parasite forms in the peripheral blood
consider: fever A03

Summary of main changes to components 1 and 7 from ICPC-1 to ICPC-2

Only major changes are listed here; additions, change in meaning of the rubric, or transfer or deletion of a rubric. There are many other changes of detail to the titles of the rubrics which do not change the meaning, and are not listed here.

CODE	TITLE ICPC-1 (some abbreviated)	TITLE ICPC-2
A05	GENERAL DETERIORATION	FEELING ILL
A11	(omitted by mistake from ICPC)	CHEST PAIN NOS
A12	ALLERGY/ALLERGIC REACTION	(transferred to A92)
A13	CONCERN ABOUT DRUG REACTION	CONCERN/FEAR ABOUT TREATMENT
A14	INFANTILE COLIC	(deleted, included in D01)
A15	EXCESSIVE CRYING INFANT	(deleted, included in A16)
A17	OTHER GEN SYMPT INFANT	(deleted, included in A16)
A18	(new rubric in ICPC-2)	CONCERN ABOUT APPEARANCE
A21	(new rubric in ICPC-2)	RISK FACTOR FOR MALIGNANCY
A23	(new rubric in ICPC-2)	RISK FACTOR NOS
A92	TOXOPLASMOSIS (deleted, included with A78)	ALLERGY/ALLERGIC REACTION NOS (transferred from A12)
A98	(new rubric in ICPC-2)	HEALTH MAINTENANCE/ PREVENTIVE MEDICINE
B03	OTHER SYMPT LYMPH GLANDS	(deleted, included in B02)
B85	UNEXPLAINED ABNORMAL BLOOD TEST	(deleted, included in A91)
B86	OTHER HAEMATOLOGICAL ABNORMALITY	(deleted, included in B99)
D07	(new rubric in ICPC-2)	DYSPEPSIA/INDIGESTION
D22	WORMS/PINWORMS/OTHER PARASITES	(transferred to D96)
D23	(transferred from D96)	HEPATOMEGALY
D96	HEPATOMEGALY	(transferred to D23)
D96	(changed rubric in ICPC-2)	WORMS/OTHER PARASITES
K22	(new rubric in ICPC-2)	RISK FACTOR FOR CARDIOVASCULAR DISEASE
K74	ANGINA PECTORIS	ISCHAEMIC HEART DISEASE WITH ANGINA
K76	OTHER AND CHRONIC ISCHAEMIC HEART	ISCHAEMIC HEART DISEASE, WITHOUT ANGINA
K80	ECTOPIC BEATS, ALL TYPES	CARDIAC ARRYTHMIA NOS
K81	HEART MURMER, NOS	HEART/ARTERIAL MURMER, NOS

K91	ATHEROSCLEROSIS (EXCL. HEART/BRAIN)	(included with K92 in ICPC-2)
K91	(altered rubric in ICPC-2)	CEREBROVASCULAR DISEASE
K92	OTHER ARTERIAL OBSTRUCTION	ATHEROSCLEROSIS/PERIPH VASC DIS
L05	FLANK SYMPTOMS/ COMPLAINTS	FLANK/AXILLA SYMPTOMS/ COMPLAINTS
L06	AXILLA SYMPTOMS/ COMPLAINTS	(deleted, included in L05)
L71	NEOPLASMS	MALIGNANT NEOPLASM
L83	SYNDROMES RELATED TO CERVICAL SPINE	NECK SYNDROME
L84	OSTEOARTHRITIS OF SPINE	BACK SYNDROME WITHOUT RADIATION
L86	LUMBAR DISC LESION, BACK PAIN	DISC LESION/BACK PAIN WITH RADIATION
L87	GANGLION JOINT/TENDON	BURSITIS/TENDONITIS/ SYNOVITIS NOS
L97	CHRONIC INTERNAL KNEE DERANGEM	NEOPLASM, BENIGN/UNCERTAIN
	(included with L99 in ICPC-2)	(split from L71 in ICPC-2)
N02	TENSION HEADACHE	(transferred to N95)
N08	(new rubric in ICPC-2)	ABNORMAL INVOLUNTARY MOVEMENTS (split from N06)
N80	OTHER HEAD INJURY WITHOUT SKULL FRACTURE	HEAD INJURY, OTHER
N95	(new rubric in ICPC-2)	TENSION HEADACHE (transferred from N02)
P21	OVERACTIVE CHILD, HYPERKINETIC	(transferred to P81)
P75	HYSTERICAL/ HYPOCHONDRIACAL DIS	SOMATIZATION DISORDER
P77	SUICIDE ATTEMPT	SUICIDE/SUICIDE ATTEMPT
P81	(new rubric in ICPC-2)	HYPERKINETIC DISORDER (transferred from P21)
P82	(new rubric in ICPC-2)	POST-TRAUMATIC STRESS DISORDER (split from P02)
P86	(new rubric in ICPC-2)	ANOREXIA NERVOSA, BULIMIA (transferred from T06)
R22	SYMPTOM/COMPLAINT TONSILS	(deleted, included in R21)
R70	TUBERCULOSIS	(deleted, included in A70)
R72	STREP-THROAT/SCARLET FEVER	STREP THROAT (scarlet fever included in A78)

R79	(new rubric in ICPC-2)	CHRONIC BRONCHITIS (transferred from R91)
R80	INFLUENZA WITHOUT PNEUMONIA	INFLUENZA
R82	PLEURISY	PLEURISY/PLEURAL EFFUSION (includes pleural effusion from R93)
R91	CHRONIC BRONCHITIS	(transferred to R79)
R92	(new rubric in ICPC-2)	NEOPLASM RESPIRATORY, UNCERTAIN NATURE
R93	PLEURAL EFFUSION	(deleted, included in R82)
S11	OTHER LOCALIZED SKIN INFECTION	WOUND INFECTION, POST-TRAUMATIC
S79	OTHER BENIGN NEOPLASMS OF SKIN	NEOPLASM SKIN, BENIGN/UNCERTAIN
S80	OTHER UNSPECIFIED NEOPLASM SKIN	SOLAR KERATOSIS/SUNBURN
S94	INGROWING NAIL/OTHER DISEASE OF NAIL	INGROWING NAIL (other disease of nail S99)
T06	ANOREXIA NERVOSA W/WO BULIMIA	(transferred to P86)
T15	THYROID LUMP/MASS	(deleted, included in T81)
T88	RENAL GLYCOSURIA	(deleted, included in T99)
T89	(new rubric in ICPC-2)	DIABETES, INSULIN DEPENDENT
T90	DIABETES MELLITUS	DIABETES, NON-INSULIN DEPENDENT
U08	(new rubric in ICPC-2)	URINARY RETENTION
W20	OTHER SYMPTOMS/ COMPLTS OF BREAST	(deleted, included in W19)
W21	(new rubric in ICPC-2)	CONCERN ABOUT BODY IMAGE IN PREGNANCY
W77	OTHER NON-OBSTETRICAL CONDITION	(deleted)
W85	(new rubric in ICPC-2)	GESTATIONAL DIABETES
X22	(new rubric in ICPC-2)	CONCERN ABOUT BREAST APPEARANCE
X92	(new rubric in ICPC-2)	CHLAMYDIA INFECTION, GENITAL

A General and unspecified

Component 1—Symptoms and complaints

Note: In this classification general or multiple refers to three or more body sites or systems. Conditions affecting one or two sites should be coded to the appropriate sites.

A01 Pain, general/multiple sites *R52*

incl: chronic general pain, multiple aches
excl: pain unspecified A29

A02 Chills *R50, R68.8*

incl: rigors, shivers
excl: fever A03

A03 Fever *R50*

incl: pyrexia
excl: fever with rash A76, heat exhaustion or stroke A88

A04 Weakness/tiredness general *G93.3, R53*

incl: chronic fatigue syndrome, exhaustion, fatigue, lassitude, lethargy, postviral fatigue
excl: feeling ill, malaise A05; drowsy A29; heat exhaustion A88; jetlag A88; somnolence P06

A05 Feeling ill *R53*

incl: malaise
excl: senescence, senility P05, cachexia T08, malnutrition T91

A06 Fainting/syncope *R55*

incl: blackout, collapse, vasovagal attack
excl: coma A07; feeling faint/giddiness/dizziness N17

A07 Coma *R40*

incl: stupor
excl: syncope A06

A08 Swelling *R68.8*

incl: lump, mass NOS
excl: lymph node B02, oedema KO7, joint L20, breast X19

A09 Sweating problems *R61*

incl: hyperhidrosis, night sweats, perspiration problems
excl: sweat gland disease S92

A10 Bleeding, haemorrhage NOS *R58*

A11 Chest pain NOS ***R07.3, R07.4***

excl: pain attributed to heart K01, to chest wall L04, to respiratory system R01

(A12) transferred to A92

A13 Concern/fear about treatment ***Z71.1***

incl: concern about the consequences of drug or any kind of treatment
excl: adverse effect of drug A85, complication of medical or surgical treatment A87

(A14) transferred to D01

(A15) deleted, included with A16

A16 Irritable infant ***R68.1***

incl: excessive crying, restless infant
excl: infantile colic D01; restless child or adult P04

(A17) deleted, included with A16

A18 Concern about appearance ***R46.8***

incl: concern about body image
excl: concern about appearance of ears H15, related to pregnancy W21, breasts X22

A20 Euthanasia request/discussion ***Z71.8***

A21 Risk factor for malignancy ***Z80, Z85***

incl: personal or family history, past treatment, or other risk factor for malignancy

A23 Risk factor NOS

Z20, Z23 to Z29, Z72.0 to Z72.5, Z73.2, Z81, Z82.0 to Z82.2, Z82.5 to Z82.8, Z83, Z84, Z86.0 to Z86.6, Z87, Z88, Z91, Z92

incl: contact with infectious disease, personal or family history, previous episode, or other risk factor for other diseases
excl: risk factor for malignancy A21, for cardiovascular disease K22

A25 Fear of death, dying ***Z71.1***

A26 Fear of cancer, NOS ***Z71.1***

excl: if patient has cancer code the disease
criteria: concern or fear of cancer in a patient without the disease or until the diagnosis is proven

A27 Fear of other disease NOS ***Z71.1***

excl: if patient has the disease code the disease (*cont.*)

criteria: concern or fear of other disease in a patient without the disease or until the diagnosis is proven

A28 Limited function/disability NOS

Z73.6, Z74, Z93.8, Z93.9, Z99.0, Z99.3, Z99.8, Z99.9

incl: limitation of function or disability not related to a disease in any other chapter
excl: falls A29
Note: The COOP/WONCA Charts are suitable for documenting the patient's functional status (see Chapter 8).

A29 General symptom/complaint, other ***R68.0, R68.8***

incl: clumsiness, drowsy, falls, pain unspecified

Component 7—Diagnoses/diseases

A70 Tuberculosis ***A15 to A19, B90***

incl: all tuberculosis infections of any body site, late effects
criteria: conversion to a positive tuberculin skin test; or demonstration of *Mycobacterium tuberculosis* or *M. bovis* on microscopy or culture; or characteristic chest X-ray appearance; or characteristic histological appearance on biopsy
consider: fever A03, cough R05

A71 Measles ***B05***

incl: complications of measles
criteria: prodrome with injected conjunctivae, fever, and cough; plus white specks on a red base in the mucous membranes of the cheek (Koplik's spots), or confluent maculopapular eruption spreading over the face and body, or an atypical exanthem in a partially immune person during an epidemic of measles; or serological evidence of acute measles
consider: fever A03, viral disease with exanthem A76, rash S07

A72 Chickenpox ***B01***

incl: complications of chickenpox
excl: herpes zoster S70
criteria: a vesicular exanthem which appears in successive crops, with the lesions evolving rapidly from superficial papules to vesicles and eventually to scabs
consider: fever A03, viral disease with exanthem A76, rash S07

A73 Malaria ***B50 to B54***

incl: complications of malaria
excl: malaria prophylaxis A44

(*cont.*)

criteria: intermittent fever with chills and rigors in a resident of, or recent visitor to, a malarial region; or demonstration of malarial parasite forms in the peripheral blood
consider: fever A03

A74 Rubella *B06, P35.0*

incl: complications of rubella, congenital rubella
excl: roseola infantum A76
criteria: an acute exanthem with enlarged lymph nodes, most often suboccipital and post-auricular, with a macular rash on the face, spreading to the trunk and proximal portions of the limbs; or serological evidence of rubella infection
consider: fever A03, viral disease with exanthem A76, rash S07

A75 Infectious mononucleosis *B27*

incl: glandular fever
criteria: inflammation of the tonsils or pharynx with lymphadenopathy not confined to the anterior cervical nodes, and either atypical lymphocytes on blood smear or splenomegaly; or abnormal heterophile antibody titre or Epstein-Barr virus titre
consider: fever A03; enlarged lymph nodes B02; acute upper respiratory tract infection R74

A76 Viral exanthem, other *A88.0, B03, B04, B08.0, B08.2 to B08.4, B08.8, B09*

incl: fever with rash, fifth disease, roseola infantum
excl: diseases specified in A71, A72, A74, A75

A77 Viral disease, other/NOS
A82, A90 to A96, A98, A99, B00.7, B25, B33.0, B33.1, B33.3, B33.8, B34, B97

incl: adenovirus; cowpox; Coxsackie diseases; dengue fever; hand foot and mouth disease; Ross River fever
excl: viral exanthems A76, influenza R80

A78 Infectious disease, other/NOS
A20 to A28, A30, A31, A32.7 to A32.9, A38, A39.1 to A39.9, A40 to A44, A48.0, A48.2 to A48.4, A48.8, A49, A59.8, A59.9, A64, A68, A69.2, A69.8, A70, A74.8, A74.9, A75, A77 to A79, B37.7 to B37.9, B38 to B49, B55, B57, B58.8, B58.9, B59, B60, B64, B89, B92, B94.8, B94.9, B95, B96, B99

incl: brucellosis; infections with site not specified; Lyme disease; meningoccocal infections; mycoplasma; Q fever; rickettsial disease; scarlet fever; sexually transmitted disease NOS; thrush NOS; toxoplasmosis
excl: meningococcal meningitis N71

A79 Maligancy NOS

C38.1 to C38.3, C38.8, C45.7, C45.9, C46.7 to C46.9, C76, C78, C79, C80, C97, D09.7, D09.9

incl: secondary and metastatic neoplasms where primary site is unknown, carcinomatosis (unknown primary)
criteria: histological evidence of malignancy
consider: other generalized disease A99

A80 Trauma/injury, NOS

S11, S15, S21, S25, S26, S27, S35, S37.9, S38.1, S38.3, S39, S45, S55, S65, S75, S85, S95, T09, T14.5, T14.7 to T14.9, T28.4, T28.9

incl: road traffic accident
excl: multiple trauma A81, late effects of trauma A82

A81 Multiple trauma/injuries

S31.7, S36.7 to S36.9, S37.7, S37.8, S39, T00 to T05, T06.5 to T06.8, T07, T29

incl: internal injuries NOS

A82 Secondary effect of trauma

T79.0 to T79.2, T79.4, T79.5, T79.7 to T79.9, T90 to T98

incl: deformities, scarring resulting from previous injury; old amputation
excl: effects related to specific body systems (code to system chapter); psychological effects P02; acute stress reaction P02; post-traumatic stress disorder P82; wound infection S11; scar of skin S99

A84 Poisoning by medical agent ***T36 to T50***

incl: drugs of addiction, overdose
excl: medication abuse P18; suicide attempt P77; insulin coma T87
criteria: toxicity or impairment produced by accidental or deliberate overdose of an agent which has remedial properties in its usual dosage

A85 Adverse effect medical agent

D61.1, D64.2, G44.4, I95.2, L27.0, L27.1, T88.6, T88.7

incl: side effects, allergy and anaphylaxis due to medication in proper dose
excl: overdose or poisoning A84; reaction to immunization and transfusion A87; Parkinsonism N87; addiction or habituation P18; contact dermatitis S88; insulin coma T87
criteria: symptoms and complaints attributed to the proper use of medication, rather than due to disease or injury

Note: May also code the nature of the adverse effect.

A86 Toxic effect non-medicinal substance *D61.2, D64.2, T51 to T65*

incl: carbon monoxide; general or local effects; industrial materials; lead; poisonous animals, insects, plants, snakes

excl: due to medical agents A84, A85, or to alcohol, tobacco, drugs P16 to P19; respiratory toxic effects R99; chemical burns—internal or external S14; non-toxic bites S12, S13; contact dermatitis S88

A87 Complication of treatment

E89, G97, H95, I97, J95, K91.0, K91.3, M96, N99, O29, O74, O86.0, O89, O90.0 to O90.2, T80, T81, T86, T87

incl: anaesthetic shock; immunization and transfusion reactions; post-operative wound disruption, infection, haemorrhage; problems due to radiation for diagnosis or treatment

excl: overdose medication A84; adverse effects of medication A85

criteria: an unexpected disorder resulting from surgical, medical, or X-ray treatment, or any other medical management

A88 Adverse effects physical factors

T33, T34, T35, T66 to T69, T70.2 to T70.4, T70.8, T70.9, T71, T73, T75, T78.8, T78.9

incl: cold; chilblains; drowning; heat; jetlag; lightning; motion; pressure; radiation

excl: effects of medical radiation A87; snow blindness F79; effects of alcohol P16; effects of tobacco P17; burns due to radiation S14; sunburn S80

A89 Effects prosthetic device *K91.4, T82 to T85, Z95 to Z97*

incl: disability, discomfort, handicap, pain or limitation of function resulting from the fitting and wearing of a device for supplying or amending deficiencies; colostomy; catheter; gastrostomy; heart valve; joint replacement; organ transplant; pacemaker

excl: dentures, false teeth D19

A90 Congenital anomaly NOS/multiple

Q85 to Q87, Q89.3, Q89.4, Q89.7, Q89.9, Q90 to Q93, Q95 to Q99

incl: Down's syndrome; Marfan's syndrome; other chromosome abnormalities

excl: anomalies related to specific body system (code to system chapter), congenital rubella A94

A91 Abnormal results investigation NOS

E83, R73, R74, R76 to R79, R83 to R94

incl: abnormal unexplained pathology or imaging test; electrolyte disorder, hyperglycaemia

(*cont.*)

excl: abnormal white cells B84; haematological abnormality B99; abnormal urine test U98; abnormal Pap smear X86
criteria: abnormal results not attributed to known disease

A92 Allergy/allergic reaction NOS *T78.0, T78.4*

incl: all allergic reactions not specified elsewhere, allergic oedema, anaphylactic shock, angioneurotic oedema
excl: urticaria S98, resulting from medication A85, allergic rhinitis R97

A93 Premature newborn *P07*

criteria: livebirth under 37 weeks' gestation

A94 Perinatal morbidity, other

P00 to P05, P08, P10 to P15, P20 to P29, P35 to P39, P50 to P61, P70 to P72, P74, P76 to P78, P80, P81, P83, P90 to P94, P96

excl: congenital conditions A90; failure to thrive T10
criteria: morbidity originating in utero or within 7 days of birth

A95 Perinatal mortality *P95, R95*

criteria: death in utero or within 7 days of birth

A96 Death *R95 to R99*

excl: perinatal A95

A97 No disease *Z00*

incl: no illness or disease detected at encounter

A98 Health maintenance/preventive medicine *Z51.8*

A99 General disease NOS

D15.7, D15.9, D36.7, D36.9, D48.9, R69, Z22, Z90.0, Z90.8, Z94.8, Z94.9, Z98.8

incl: disease or condition of unspecified nature or site; disease carrier

B Blood, blood forming organs, and immune mechanism

Component 1—Symptoms and complaints

B02 Lymph gland(s) enlarged/painful *R59*

incl: lymphadenopathy with or without pain/tenderness

(*cont.*)

excl: lymphadenitis—acute B70, chronic B71

(B03) deleted, included with B02

B04 Blood symptom/complaint *R68.8*

excl: anaemia B82, pallor S08

B25 Fear of AIDS *Z71.1*

excl: if patient has the disease code the disease
criteria: concern or fear of AIDS or HIV in a patient without the disease or until the diagnosis is proven

B26 Fear of cancer blood/lymph *Z71.1*

excl: if patient has the disease code the disease
criteria: concern or fear of these diseases in a patient without the disease or until the diagnosis is proven

B27 Fear of blood/lymph disease, other *Z71.1*

excl: if patient has the disease code the disease
criteria: concern or fear of other disease in a patient without the disease or until the diagnosis is proven

B28 Limited function/disability blood/lymph *Z73.6*

incl: disability due to bleeding disorders
Note: The COOP/WONCA Charts are suitable for documenting the patient's functional status (see Chapter 8).

B29 Sympt/complt blood and immune mechanism *R68.8*

excl: splenomegaly B87

Component 7—Diagnoses/diseases

B70 Lymphadenitis, acute *L04*

incl: abscess of lymph node
excl: chronic lymphadenitis B71; mesenteric lymphadenitis B71; acute lymphangitis S76
criteria: one or more inflamed or enlarged and tender or painful lymph nodes in the same anatomical location, which are of recent onset (less than 6 weeks) and where the primary source of infection is unknown
consider: enlarged lymph node B02

B71 Lymphadenitis, non-specific *I88*

incl: mesenteric lymphadenitis, acute or chronic
excl: acute lymphadenitis (except mesenteric) B70, acute lymphangitis S76
criteria: enlarged tender lymph nodes present for more than 6 weeks, or demonstration of enlarged inflamed mesenteric lymph nodes by surgery, sonography, lymphography, or otherwise
consider: enlarged lymph node B02

B72 Hodgkins's disease/lymphomas *C81 to C85*

criteria: characteristic histologic appearance
consider: other malignant neoplasms B74; neoplasms not specified benign or malignant (when histology is not available) B75

B73 Leukaemia *C91 to C95*

incl: all types of leukaemia
criteria: characteristic histologic appearance
consider: neoplasms not specified as benign or malignant (when histology is not available) B75

B74 Malignant neoplasm blood, other *C37, C46.3, C77, C88, C90, C96*

incl: myeloproliferative disease, multiple myeloma
excl: Hodgkin's disease, lymphomas B72

B75 Benign/uncertain neoplasms blood *D15.0, D36.0, D45, D47*

incl: polycythaemia rubra vera

B76 Ruptured spleen, traumatic *S36.0*

B77 Injury blood/lymph/spleen other *T14.9*

B78 Hereditary haemolytic anaemias *D56 to D58*

incl: sickle-cell anaemia, sickle-cell trait, spherocytosis, thalassaemia
criteria: characteristic findings by tests such as haemoglobin electrophoresis, blood smear, or increased osmotic fragility of red cells
consider: other congenital anomalies B79

B79 Congenital anomaly blood/lymph, other
D61.0, D64.0, D64.4, Q89.0, Q89.8

incl: congenital anaemia
excl: hereditary haemolytic anaemia B78, haemophilia B83, haemangioma S81, lymphangioma S81

B80 Iron deficiency anaemia *D50*

incl: anaemia due to blood loss
excl: iron deficiency without anaemia T91
criteria: decrease in haemoglobin or haematocrit below levels appropriate for age and sex; plus evidence of blood loss, or microcytic hypochromic red cells by appearance or indices in the absence of thalassaemia, or decreased serum iron and increased iron-binding capacity, or decreased serum ferritin, or reduced haemosiderin in bone marrow, or good response to iron administration
consider: anaemia unspecified B82

B81 Anaemia, Vit B12/folate deficiency *D51, D52*

incl: macrocytic anaemia, pernicious anaemia
excl: vitamin B12 deficiency without anaemia T91
criteria: macrocytic anaemia by smear or indices; plus decreased vitamin B12 level, folate level, or positive Schilling test

B82 Anaemia other/unspecified

D46, D53, D55, D59 to D60, D61.3, D61.8, D61.9, D62, D64.1, D64.3, D64.8, D64.9

incl: aplastic anaemia, acquired haemolytic anaemia, megaloblastic anaemia NOS, protein deficiency anaemia
excl: iron deficiency anaemia B80; vitamin B12/folate deficiency anaemia B81; anaemia of pregnancy W84

B83 Purpura/coagulation defects *D65 to D69*

incl: abnormal platelets, haemophilia, thrombocytopenia

B84 Abnormal white cells *D70 to D72, R72*

incl: agranulocytosis; eosinophilia; leukocytosis; lymphocytosis; neutropenia
excl: leukaemia B73

(B85) deleted, included with A91

(B86) deleted, included with B99

B87 Splenomegaly *R16.1, R16.2*

excl: hypersplenism B99

B90 HIV infection, AIDS *B20 to B24, R75, Z21*

B99 Blood/lymph/spleen disease, other

D73 to D76, D80 to D84, D86, D89, I89, R70, R71

incl: autoimmune disease of blood; complement defects; hypersplenism; immunodeficiency disorder; other or unspecified haematological abnormality; raised ESR; red cell abnormality; sarcoidosis; secondary polycythaemia
excl: lymphadenitis B70, B71; primary polycythaemia B75; human immunodeficiency virus (HIV) disease B90; lymphoedema K99

D Digestive

Component 1—Symptoms and complaints

D01 Abdominal pain/cramps, general ***R10.0, R10.4***

incl: abdominal colic, cramps, discomfort, pain NOS; infant colic
excl: stomach ache D02, heartburn D03, localized abdominal pain D06, indigestion D07, gas pain D08, biliary colic D98, renal colic U14, dysmenorrhoea X02

D02 Abdominal pain, epigastric ***R10.1***

incl: epigastric discomfort, fullness; stomach ache or pain
excl: dyspepsia/indigestion D07; flatulence, gas, wind D08

D03 Heartburn ***R12***

incl: acidity, waterbrash
excl: epigastric pain D02; indigestion D07; oesophagitis, reflux D84

D04 Rectal/anal pain ***K59.4, R10.2, R10.3***

incl: anal spasm; proctalgia fugax
excl: impacted faeces D12

D05 Perianal itching ***L29.0, L29.3***

incl: perianal pruritis
excl: pruritis vulvae X16

D06 Abdominal pain, localized, other ***R10.1 to R10.3***

incl: colonic pain
excl: epigastric pain D02

D07 Dyspepsia/indigestion ***K30***

excl: epigastric pain D02, heartburn D03, belching D08

D08 Flatulence/gas/belching *R14*

incl: bloating; eructation; gas pains; gaseous distension; passing wind
excl: indigestion NOS D07, change in abdominal size D25

D09 Nausea *R11*

excl: feelings of over-eating D02; vomiting D10; alcohol induced P16; loss of appetite T03; pregnancy W05
Note: code nausea and vomiting in problem mode as D10.

D10 Vomiting *F50.5, R11*

incl: emesis, hyperemesis, retching
excl: haematemesis D14; in pregnancy W05
Note: code diarrhoea and vomiting in problem mode as D11.

D11 Diarrhoea *K52.9, K59.1*

incl: frequent or loose bowel movements, watery stools
excl: melaena D15, other bowel changes D18

D12 Constipation *K56.4, K59.0*

incl: faecal impaction
excl: ileus D99

D13 Jaundice *R17*

incl: icterus

D14 Haematemesis/vomiting blood *K92.0*

excl: haemoptysis R24

D15 Melaena *K92.1*

incl: black, tarry stools
excl: fresh blood in stool D16

D16 Rectal bleeding *K62.5*

incl: fresh blood in stool
excl: melaena D15

D17 Incontinence of bowel *R15*

incl: faecal incontinence
excl: encopresis P13

D18 Change in faeces/bowel movements *R19.4, R19.5*

excl: diarrhoea D11, constipation D12, incontinence D17

D19 Teeth/gum symptom/complaint *K00.7*

incl: denture problem, gingival inflammation or bleeding, teething, toothache
excl: caries D82

D20 Mouth/tongue/lip symptom/complaint *K14, R19.6, R68.2*

incl: bad breath; coated tongue; cracked lips; dribbling; dry mouth; halitosis; swollen lips; sore mouth
excl: cheilosis D83; dental and gum problems D19; abnormal taste N16; dehydration T11

D21 Swallowing problems *R13*

incl: choking feeling, dysphagia

(D22) (transferred to D96)

D23 Hepatomegaly *R16.0, R16.2*

D24 Abdominal mass NOS *R19.0*

incl: lump abdomen
excl: splenomegaly B87, hepatomegaly D23, renal mass U14

D25 Abdominal distenstion *R19.0*

incl: abdominal swelling without mass
excl: bloating, gas D08; abdominal mass D24; ascites D29

D26 Fear of cancer of digestive system *Z71.1*

excl: if the patient has the disease code the disease
criteria: concern or fear of cancer of digestive system in a patient without cancer or until the diagnosis is proven

D27 Fear of digestive disease, other *Z71.1*

incl: fear of digestive disease other than cancer
excl: if the patient has the disease code the disease
criteria: concern or fear of other digestive disease in a patient without the disease or until the diagnosis is proven

D28 Limited function/disability digestive *Z73.6, Z93.1 to Z93.4*

excl: colostomy, gastrostomy A89; post-surgery disorders D99; dumping syndrome D99

Note: The COOP/WONCA Charts are suitable for documenting the patient's functional status (see Chapter 8).

D29 Digestive symptom/complaint, other *R18, R19.1 to R19.3, R19.8*

incl: ascites, teeth grinding

Component 7—Diagnosis/diseases

D70 Gastrointestinal infection *A00 to A08*

incl: gastrointestinal infection or dysentery with specified organisms including campylobacter, giardia, salmonella, shigella, typhoid, cholera

excl: contacts, carriers of infective/parasitic disease A99; presumed or unspecified infections D73

criteria: a symptomatic patient with isolation or serological evidence of pathogenic bacteria, virus, or protozoan from either the stool or from food ingested

consider: gastroenteritis presumed infection D73

D71 Mumps *B26*

incl: mumps meningitis, mumps orchitis, mumps pancreatitis

criteria: acute non-suppurative, non-erythematous diffuse tender inflammation of one or more salivary glands; or acute mumps infection demonstrated by culture or serology; or orchitis in a person exposed to mumps following appropriate incubation period

consider: mass or localized swelling A08

D72 Viral hepatitis *B15 to B19*

incl: all hepatitis presumed viral, chronic active hepatitis

excl: hepatitis NOS D97

criteria: evidence of viral infection with inflammation of the liver with or without jaundice; or serological evidence of an infection with a hepatitis virus

consider: jaundice D13, hepatomegaly D23

D73 Gastroenteritis, presumed infection *A09*

incl: diarrhoea or vomiting presumed to be infective; dysentery NOS; food poisoning; gastric flu

excl: functional digestive disorders D93; non-infective enteritis and gastroenteritis D94, D99

D74 Malignant neoplasm stomach *C16*

incl: carcinoma stomach
criteria: characteristic histological appearance
consider: other malignant neoplasms (when primary site is uncertain) D77; neoplasms uncertain whether benign or malignant (when histology is not available) D78

D75 Malignant neoplasm colon/rectum *C18 to C21*

criteria: characteristic histological appearance
consider: other malignant neoplasms (when primary site is uncertain) D77; neoplasms uncertain whether benign or malignant (when histology is not available) D78

D76 Malignant neoplasm pancreas *C25*

incl: carcinoma pancreas
criteria: characteristic histological appearance
consider: neoplasms uncertain whether benign or malignant (when histology is not available) D78

D77 Malignant neoplasm digestive, other/NOS

C00 to C08, C14.8, C15, C17, C22 to C24, C26, C45.1, C46.2, C48

incl: all primary malignancies of digestive system except D74 to D76; gallbladder cancer; liver cancer
excl: secondary malignancy of known site (code to site); unknown site A79

D78 Neoplasm digestive, benign/uncertain

D00, D01, D10 to D13, D20, D37, D48.3, D48.4, K62.0, K62.1

incl: polyps of stomach, duodenum, colon, rectum

D79 Foreign body digestive system *T18*

incl: foreign body in digestive tract including mouth, oesophagus, swallowed, rectum
excl: foreign body in throat R87, inhaled R87

D80 Injury, other

S00.5, S01.5, S02.5, S03.2, S36.1 to S36.6, T28.0 to T28.2, T28.5 to T28.7

incl: injury to abdominal organ, teeth, tongue
excl: multiple organ injuries A81, pelvic organs X82, Y80

D81 Congenital anomaly digestive *Q18, Q35 to Q45*

incl: biliary anomalies; cleft lip and palate; Meckel's diverticulum; megacolon, Hirschprung's disease; oesophageal atresia; pyloric stenosis; tongue-tie

(*cont.*)

excl: haemangioma, lymphangioma S81; congenital metabolic disorders T80

D82 Teeth/gum disease — *K00.0 to K00.6, K00.8, K00.9, K01 to K10*

incl: caries; dental abscess; gingivitis; malocclusion; temporomandibular joint disorders

excl: teething D19; denture problem D19; injuries to teeth and gums D80; Vincent's angina D83

D83 Mouth/tongue/lip disease — *A69.0, A69.1, B00.2, B37.0, K11 to K14*

incl: aphthous ulcer; cheilosis; glossitis; mucocoele; parotitis; oral thrush; salivary calculus; stomatitis; Vincent's angina

excl: mumps D71; injuries D80; herpes simplex S71

D84 Oesophagus disease — *K20 to K22*

incl: achalasia, diverticulum, Mallory–Weiss syndrome, oesophagitis, reflux, ulceration

excl: cancer of oesophagus D77; hiatus hernia D90; oesophageal varices K99

D85 Duodenal ulcer — *K26*

incl: bleeding, obstructing, perforated

criteria: characteristic imaging findings; or characteristic endoscopy findings; or exacerbation of symptoms in a patient with a previously proven duodenal ulcer

consider: heartburn D03; indigestion D07

D86 Peptic ulcers, other — *K25, K27, K28*

incl: gastric, gastrojejunal, marginal; acute erosions; peptic ulcer NOS

excl: oesophageal ulcer D84; duodenal ulcer D85

criteria: characteristic imaging findings; or characteristic endoscopy findings; or exacerbation of symptoms in a patient with a previously proven ulcer

consider: heartburn D03; indigestion D07

D87 Stomach function disorder — *K29*

incl: acute dilatation stomach; duodenitis; gastritis (incl alcoholic)

excl: infective gastritis or enteritis D70, D73

criteria: disorder of stomach function proven by investigation

consider: abdominal pain D01, D06; epigastric pain D02; heartburn D03; indigestion/dyspepsia D07; gas problems (wind) D08; nausea D09; oesophagitis D84; vomiting D10

D88 Appendicitis — *K35 to K37*

incl: appendix abscess, perforation

(*cont.*)

criteria: objective evidence of inflammation of the appendix, such as demonstrated at operation or pathological examination
consider: abdominal pain D01, D06; vomiting D10

D89 Inguinal hernia *K40*

excl: femoral hernia D91
criteria: swelling in the inguinal region and transmitted impulse with cough; or enlargement on straining; or reducible into the abdomen; or intestinal obstruction
consider: abdominal lump D24

D90 Hiatus hernia *K44*

incl: diaphragmatic hernia
excl: oesophagitis, reflux D84
criteria: characteristic findings on imaging, endoscopy, intraluminal pressure studies; or at surgery
consider: epigastric pain D02; heartburn D03; indigestion D07

D91 Abdominal hernia, other *K41 to K43, K45, K46*

incl: femoral, incisional, umbilical, ventral
excl: hiatus hernia D90
criteria: demonstration at surgery; or swelling in the specified area and transmitted impulse with cough, or enlargement on straining, or reducible into the abdomen, or intestinal obstruction
consider: abdominal lump D24

D92 Diverticular disease *K57*

incl: diverticulitis; diverticulosis of intestine
excl: Meckel's diverticulum D81; oesophageal diverticulum D84
criteria: imaging demonstration of diverticula; or demonstration of diverticula at surgery; or acute abdominal pain with fever and palpable tender descending or sigmoid colon
consider: abdominal pain D01, D06

D93 Irritable bowel syndrome *K58*

incl: mucous colitis, spastic colon
excl: intestinal disease of infective origin D70, D73; regional enteritis D94; allergic, dietetic, and toxic gastroenteritis and colitis D99; vascular insufficiency of gut D99; psychogenic diarrhoea P75
criteria: continuous or intermittent abdominal pain and variable bowel pattern over a period of time; and increased gas, or tender and palpable colon, or history of mucous without blood in stool
consider: abdominal pain D01, D06; flatulence D08; diarrhoea D11; constipation D12

D94 Chronic enteritis/ulcerative colitis *K50, K51, K52.0*

incl: Crohn's disease, regional enteritis, ulcerative colitis
criteria: characteristic endoscopic, imaging, or histological findings for these conditions
consider: abdominal pain D01, D06; diarrhoea D11; mucus colitis D93

D95 Anal fissure/perianal abscess *K60, K61*

incl: anal fistula, ischiorectal abscess
excl: pilonidal abscess S85

D96 Worms/other parasites *B65 to B83*

incl: cestodes; creeping eruption; intestinal parasites unspecified; trichiniasis; hydatid disease
criteria: either demonstration of helminth in adult form, larvae, or ova; or positive skin tests; or positive serology

D97 Liver disease NOS *B58.1, B94.2, K70 to K76*

incl: alcohol hepatitis, cirrhosis, fatty liver, hepatitis NOS, liver failure, portal hypertension
excl: viral hepatitis D72, hydatid disease D96

D98 Cholecystitis, cholelithiasis *K80 to K83*

incl: biliary colic, cholangitis, gallstones
criteria: cholecystitis: demonstration of typical pathology by ultrasound or surgery; or localized right upper quadrant tenderness and jaundice or fever or history of gallstones
cholelithiasis: imaging or surgical demonstration of gallstones
acute biliary colic: acute colicky right upper quadrant abdominal pain without fever; and jaundice or right upper quadrant abdominal tenderness, or history of gallstones
consider: localized abdominal pain D06

D99 Disease digestive system, other

K31, K38, K52.1, K52.2, K52.8, K52.9, K55, K56.0 to K56.3, K56.5 to K56.7, K59.2, K59.3, K59.8, K59.9, K62.2 to K62.4, K62.6 to K62.9, K63 to K66, K85 to K86, K90, K91.1, K91.2, K91.5 to K91.9, K92.2, K92.8, K92.9, Z90.3, Z90.4, Z94.4, Z98.0

incl: abdominal adhesions; coeliac disease; dumping syndrome; food intolerance; gastroenteropathy allergic, toxic, and dietetic; ileus; intestinal obstruction; intussuception; malabsorption syndrome; mesenteric vascular disease; pancreatic diseases; peritonitis; secondary megacolon; sprue; and other disease of the digestive system NOS
excl: antibiotic associated colitis A85, malignant disease D74–D77

F Eye

Component 1—Symptoms and complaints

F01 Eye pain *H57.1*

excl: abnormal sensations in eye F13

F02 Red eye *H57.8*

incl: bloodshot eye, inflamed eye

F03 Eye discharge *H04.2*

incl: lacrimation, purulent discharge, watery eye

F04 Visual floaters/spots *H53.8*

incl: spots in the visual field, fixed or floating
excl: visual disturbance F05

F05 Visual disturbance, other *H53.1 to H53.3, H53.9, H54.7*

incl: blurred vision; difficulty reading; diplopia; eye strain; photophobia; scotoma and dazzle when symptoms confined to eyes; temporary blindness NOS; weak eyes
excl: blindness one eye F28, snowblindness F79, refractive errors F91, permanent blindness F94, colour blindness F99, night blindness F99

F13 Eye sensations abnormal *H57.8*

incl: burning eye; dry eye; itchy eye
excl: eye pain F01

F14 Eye movements abnormal *H55*

incl: abnormal blinking, lazy eye, nystagmus
excl: squint F95; twitching N08; tic of eye P10

F15 Eye appearance abnormal *H57.8*

incl: change eye colour; swollen eyes
excl: red eye F02

F16 Eyelid symptom/complaint *H02.2 to H02.7*

incl: ptosis
excl: inflamed eyelids F72

F17 Glasses symptom/complaint *Z46.0*

incl: problems from spectacles affecting structure, function, or sensations of eyes
excl: problems with contact lens F18

F18 Contact lens symptom/complaint *Z46.0*

incl: problems from contact lens affecting structure, function, or sensations of eyes

F27 Fear of eye disease *Z71.1*

incl: fear of blindness
excl: if the patient has the disease code the disease
criteria: concern or fear of eye disease in a patient without the disease or until the diagnosis is proven

F28 Limited function/disability eye *H54.4 to H54.6, Z73.6*

incl: blindness one eye; disability due to eye or vision problem
excl: blindness F94
Note: The COOP/WONCA Charts are suitable for documenting the patient's functional status (see Chapter 8).

F29 Eye symptom/complaint, other *H57.9*

Component 7—Diagnosis/diseases

F70 Conjunctivitis, infectious

A74.0, B30, H10.0, H10.2 to H10.5, H10.8, H10.9

incl: bacterial and viral and NOS
excl: allergic with and without rhinorrhoea F71, flash burn F79, trachoma F86
criteria: inflammation of conjunctivae in the absence of those conditions listed under exclusions

F71 Conjuctivitis, allergic *H10.1*

incl: allergic with or without rhinorrhea
excl: bacterial and viral conjunctivitis F70, flash burn F79, trachoma F86
criteria: hyperaemia of conjunctivae, excess watering of eyes, itching and possible oedema of conjunctivae in absence of those conditions listed under exclusions

F72 Blepharitis/stye/chalazion *H00, H01*

incl: dermatitis; dermatosis of eyelids; hordeolum; infection; meibomian cyst; tarsal cyst

(*cont.*)

excl: dacrocystitis F73
criteria: generalized or localized inflammation and/or swelling of eyelids including the tarsal glands

F73 Eye infections/inflammation, other ***B00.5, B58.0, H04.3, H04.4, H05.1, H05.2, H16.1 to H16.4, H16.8, H16.9, H20 to H21, H30***

incl: dacryocystitis; herpes simplex without corneal ulcer; inflammation of the orbit; iritis; iridocyclitis; keratitis
excl: measles keratitis A71, corneal ulcer (herpes) F85, trachoma F86, herpes zoster S70

F74 Neoplasm of eye/adnexa ***C69, D09.2, D31, D48.7***

incl: benign and malignant

F75 Contusion/haemorrhage eye ***H11.3, S00.1, S05.1***

incl: black eye; hyphaema; subconjunctival haemorrhage
excl: corneal ulcer F85

F76 Foreign body in eye ***T15***

excl: corneal abrasion F79

F79 Injury eye, other ***H44.6, H44.7, S00.2, S01.1, S05.0, S05.2 to S05.9, T26***

incl: corneal abrasion, flash burn, snowblindness
excl: contusion/haemorrhage F75, foreign body F76

F80 Blocked lacrimal duct of infant ***Q10.5***

excl: dacryocystitis F73, in older person F99
criteria: overflow of tears without crying, beginning before the age of three months

F81 Congenital anomaly eye, other ***Q10.0 to Q10.4, Q10.6, Q10.7, Q11 to Q15***

F82 Detached retina ***H33***

F83 Retinopathy ***H35.0 to H35.2, H35.4, H36***

incl: diabetic, hypertensive, or other
Note: double code known causative disease eg. diabetes T89 or T90, hypertension K87.

F84 Macular degeneration ***H35.3***

excl: retinopathy F83

F85 Corneal ulcer *H16.0*

incl: dendritic ulcer; viral keratitis
excl: corneal abrasions F79, other injury F79

F86 Trachoma *A71, B94.0*

criteria: either proven infection with *Chlamydia trachomatis* or typical clinical features including chronic inflammation and hypertrophy of the conjunctiva with formation of yellowish or greyish granules
consider: red eye F02, discharge from eye F03

F91 Refractive error *H52*

incl: astigmatism; hypermetropia; long sighted; myopia; presbyopia; short sighted
excl: blindness and reduced visual acuity NOS F94
criteria: visual deficit correctible with an appropriate lens

F92 Cataract *H25, H26*

excl: congenital F81
criteria: opacity of part or all of the optic lens which reduces or impairs the vision

F93 Glaucoma *H40*

incl: raised intraocular pressure
excl: congenital F81

F94 Blindness *H54.0 to H54.3*

incl: partial or complete blindness of both eyes
excl: blurred vision F05; temporary blindness F05; blindness one eye F28; snow-blindness F79; refractive errors F91; colour blindness F99; night blindness F99

F95 Strabismus *H49 to H51*

incl: cross-eye, squint
criteria: lack of parallelism of visual axis of the eyes demonstrated at medical examination
consider: abnormal eye movement F14

F99 Eye/adnexa disease, other

H02.0, H02.1, H02.8, H02.9, H04.0, H04.1, H04.5 to H04.9, H05.2 to H05.5, H05.8, H05.9, H11.0 to H11.2, H11.4, H11.8, H11.9, H15, H17, H18, H27, H31, H34, H35.5 to H35.9, H43, H44.0 to H44.5, H44.8, H44.9, H46 to H47, H53.0, H53.4 to H53.6, H57.0, H57.8, H58, H59, Z94.7

incl: amblyopia; arcus senilis; colour blindness; corneal opacity; disorders of orbit; ectropion; entropion; episcleritis; ingrowing eyelash; nightblindness; papilloedema; pterygium; scleritis; and other diseases of the eye NOS

H Ear

Component 1—Symptoms and complaints

H01 Ear pain/earache *H92.0*

H02 Hearing complaint *H93.2*

excl: deafness one ear H28, both ears H86

H03 Tinnitus, ringing/buzzing ear *H93.1*

incl: echo in ear
excl: ears crackling or popping H29

H04 Ear discharge *H92.1*

excl: blood H05

H05 Bleeding ear *H92.2*

incl: blood in/from ear

H13 Plugged feeling ear *H93.8*

incl: blocked ear
excl: ear wax H81

H15 Concern with appearance of ears *R46.8*

excl: bat ears H80, congenital anomalies H80

H27 Fear of ear disease *Z71.1*

incl: fear of deafness
excl: in a patient with the disease code the disease
criteria: concern or fear of ear disease or deafness in a patient without the disease or until the diagnosis is excluded

H28 Limited function/disability ear *Z73.6*

incl: deafness one ear, temporary deafness
excl: deafness H84-86; dizziness, vertigo N17
Note: The COOP/WONCA Charts are suitable for documenting the patient's functional status (see Chapter 8).

H29 Ear symptom/complaint, other *H93.9*

incl: ears crackling or popping; itchy ears; pulling at ears
excl: dizziness or vertigo N17, loss of balance N17

Component 7—Diagnosis/diseases

H70 Otitis Externa *H60*

incl: eczema external auditory meatus; furuncle or abscess external auditory meatus; swimmer's ear
criteria: inflammation or desquamation of the external auditory canal

H71 Acute otitis media/myringitis *H66.0, H66.4, H66.9, H73.0*

incl: acute suppurative otitis media, otitis media NOS, acute tympanitis
excl: serous otitis media H72, chronic otitis media H74
criteria: recent perforation of the tympanic membrane discharging pus; or inflamed and bulging tympanic membrane; or one ear drum is redder than the other; or red tympanic membrane, with ear pain; or bullae on the tympanic membrane
consider: ear pain H01, ear discharge H04

H72 Serous otitis media *H65*

incl: glue ear; otitis media with effusion (OME)
excl: acute otitis media H71, chronic otitis media H74
criteria: visible fluid behind the tympanic membrane, without inflammation; or dullness of the tympanic membrane with either retraction or bulging; or dullness of the tympanic membrane with related impairment of hearing
consider: plugged feeling ear H13, eustachian salpingitis or block H73

H73 Eustachian salpingitis *H68, H69*

incl: eustachian block/catarrh/dysfunction
excl: serous otitis media H72
criteria: acute onset of impairment of hearing, or discomfort in the ear' with normal tympanic membrane
consider: plugged feeling in ear H13

H74 Chronic otitis media *H66.1 to H66.3, H70, H71, H73.1*

incl: cholesteatoma, chronic suppurative otitis media, mastoiditis
excl: serous otitis media H72

H75 Neoplasm of ear *C30.1, D48.9, H74.4*

incl: benign and malignant
excl: polyp H99, acoustic neuroma N75

H76 Foreign body in ear *T16*

H77 Perforation ear drum *H72*

excl: with infection H71, H74; traumatic or pressure rupture ear drum H79

H78 Superficial injury of ear *S00.4, S09.9*

incl: external meatus, pinna
excl: of tympanic membrane H79

H79 Ear injury, other *S01 3, S09.2, T70.0*

incl: traumatic or pressure rupture of drum

H80 Congenital anomaly of ear *Q16, Q17*

incl: accessory auricle, bat ears
excl: congenital deafness H86

H81 Excessive ear wax *H61.2*

criteria: symptoms/complaint due to wax in ear canal

H82 Vertiginous syndromes *A88.1, H81, H83.0*

incl: benign paroxysmal and positional vertigo; epidemic vertigo; labyrinthitis; Ménière's disease; vestibular neuronitis
criteria: true rotational vertigo
consider: vertigo, giddiness, dizziness N17

H83 Otosclerosis *H80*

H84 Presbyacusis *H91.1*

excl: deafness unspecified H86
criteria: gradual onset with ageing of symmetrical, bilateral deafness, particularly involving high-frequency sounds
consider: hearing impairment H28

H85 Acoustic trauma *H83.3*

incl: noise deafness
excl: perforation of ear drum H77
criteria: deafness in the high-frequency range with a definite history of exposure to loud noise
consider: deafness unspecified H86

H86 Deafness *H90, H91.0, H91.2 to H91.9*

incl: congenital deafness; partial or complete deafness both ears
excl: temporary deafness H28, deafness one ear H28, otosclerosis H83, presbyacusis H84, noise deafness H85

H99 Ear/mastoid disease, other

H61.0, H61.1, H61.3 to H61.9, H73.8, H73.9, H74, H83.1, H83.2, H83.8, H83.9, H93.0, H93.3, H93.8

incl: polyp of middle ear
excl: mastoiditis H74

K Circulatory

Component 1—Symptoms and complaints

K01 Heart pain *R07.2*

incl: pain attributed to heart by the patient
excl: chest pain NOS A11; fear of heart attack K24; angina pectoris K74; chest tightness R29

K02 Pressure/tightness of heart *R07.2*

incl: heaviness of heart according to the patient
excl: chest pain NOS A11; fear of heart attack K24; angina pectoris K74; shortness of breath, dyspnoea R02

K03 Cardiovascular pain NOS *R09.8*

excl: pain attributed to the heart K01; claudication K92, migraine N89

K04 Palpitations/awareness of heart *R00.0 to R00.2*

incl: tachycardia
excl: paroxysmal tachycardia K79

K05 Irregular heartbeat, other *R00.8*

excl: palpitations K04

K06 Prominent veins *I78.1, I87.8*

incl: any unusually prominent veins, spider naevus
excl: varicose veins K95, haemangioma S81

K07 Swollen ankles/oedema *R60*

incl: dropsy, fluid retention, swollen feet, swollen legs
excl: ankle symptoms L16, localized swelling S04

K22 Risk factor for cardiovascular disease *Z82.3, Z82.4, Z86.7*

incl: personal or family history, previous episode, or other risk factor for cardiovascular disease

K24 Fear of heart disease *Z71.1*

incl: fear of heart attack
excl: if patient has the disease code the disease
criteria: concern or fear of heart attack in a patient without a current episode of acute infarction

K25 Fear of hypertension *Z71.1*

excl: if patient has the disease code the disease
criteria: concern or fear of hypertension in a patient without the disease

K27 Fear of cardiovascular disease, other *Z71.1*

excl: if patient has the disease code the disease
criteria: concern or fear of other cardiovascular disease in a patient without the disease

K28 Limited function/disability cardiovascular *Z73.6*

excl: fear of diseases K25, K26, K27
criteria: presence of disability attributable to a symptom and/or disease of the circulatory system

Note: The COOP/WONCA Charts are suitable for documenting the patient's functional status (see Chapter 8).

K29 Cardiovascular symptom/complaint, other *R03.1, R09.8*

incl: heart trouble, low blood pressure, weak heart
excl: fluid in chest R82; cyanosis S08

Component 7—Diagnosis/diseases

K70 Infection of circulatory system *B33.2, B37.6, I30, I33, I38, I40 to I41*

incl: acute and subacute endocarditis; bacterial endocarditis; myocarditis; pericarditis (other than rheumatic)
excl: rheumatic heart disease K71; phlebitis and thrombophlebitis K94; arteritis K99

K71 Rheumatic fever/heart disease *I00 to I02, I05 to I07, I09*

incl: chorea, mitral stenosis

(cont.)

criteria: For acute rheumatic fever—two major, or one major, and two minor manifestations, plus evidence of preceding streptococcal infection:

— major manifestations: migratory polyarthritis; carditis; chorea; erythema marginatum; subcutaneous nodules of recent onset
— minor manifestations: fever; arthralgia; elevated ESR or positive C-reactive protein; prolonged P-R interval on ECG

For chronic rheumatic heart disease—either physical findings consistent with a valve lesion of the heart in a patient with a history of rheumatic fever; or physical findings consistent with mitral stenosis, even in the absence of a history of rheumatic fever, but without any other demonstrable cause

consider: heart valve disease K83, other heart disease K84

K72 Neoplasm, cardiovascular *C38.0, C45.2, D15.1, D15.2, D48.7*

incl: benign and malignant
excl: haemangioma S81
Note: code secondary and metastatic neoplasms to primary site

K73 Congenital anomaly, cardiovascular *Q20 to Q28*

incl: atrial or ventricular septal defect; Fallot's tetralogy; patent ductus arteriosus
excl: haemangioma S81

K74 Ischaemic heart disease with angina *I20, I24*

incl: angina of effort, angina pectoris, angina with spasm; ischaemic chest pain; unstable angina
excl: ischaemic heart disease without angina K76
criteria: history plus ECG or imaging evidence of old myocardial infarction; or demonstration of myocardial ischaemia by resting or exercise ECG; or investigation evidence of coronary artery narrowing or ventricular aneurysm
consider: heart pain K01

K75 Acute myocardial infarction *I21 to I23*

incl: myocardial infarction specified as acute or within 4 weeks (28 days) of onset
excl: old or healed myocardial infarct K74, K76
criteria: chest pain characteristic of myocardial ischaemia, lasting more than 15 minutes, and/or abnormal ST-T changes or new Q waves in electrocardiogram or elevation of blood cardiac enzymes
consider: heart pain K01, angina pectoris K74, chronic ischaemic heart disease K76
Note: double code K74 or K76 as well.

K76 Ischaemic heart disease without angina *I24, I25*

incl: aneurysm of heart, arteriosclerotic or atherosclerotic heart disease, coronary artery disease, ischaemic cardiomyopathy, old myocardial infarction, silent myocardial ischaemia

(*cont.*)

excl: ischaemic heart disease with angina K74
criteria: history plus ECG, or imaging evidence of old myocardial infarction; or demonstration of myocardial ischaemia by resting or exercise ECG; or investigation evidence of coronary artery narrowing; or ventricular aneurysm

K77 Heart failure *I50*

incl: cardiac asthma, congestive heart failure, heart failure NOS, left ventricular failure, pulmonary oedema, right ventricular failure
excl: cor pulmonale K82
criteria: multiple signs including dependent oedema, raised jugular venous pressure, hepatomegaly in the absence of liver disease, pulmonary congestion, pleural effusion, enlarged heart

K78 Atrial fibrillation/flutter *I48*

excl: paroxysmal tachycardia K79
criteria: demonstration of characteristic findings by electrocardiogram; or totally irregular heart rate with a pulse deficit
consider: palpitations K04, abnormal irregular heartbeat K05

K79 Paroxysmal tachycardia *I47*

incl: supraventricular or ventricular
excl: tachycardia NOS K04, atrial fibrillation K78
criteria: history of recurrent episodes of rapid heart rate (over 140/min) with both abrupt onset and termination
consider: palpitations K04, abnormal irregular heartbeat K05

K80 Cardiac arrhythmia NOS *I49*

incl: atrial, junctional and ventricular premature beats; bradycardia; bigeminy; ectopic beats; extrasystoles; premature beats; sick sinus syndrome; ventricular fibrillation and flutter
excl: paroxysmal tachycardia K79
criteria: one or more heart beats which occur at times other than the regular beats of the underlying rhythm
consider: palpitations K04, abnormal irregular heartbeat K05

K81 Heart/arterial murmur NOS *R01*

incl: cardiac bruit, carotid bruit, innocent murmur of childhood, renal artery bruit
excl: rheumatic heart disease K71, valve disease K83, cerebrovascular disease K90

K82 Pulmonary heart disease *I27, I28*

incl: chronic cor pulmonale, diseases of pulmonary vessels, primary pulmonary hypertension
excl: pulmonary embolism K93

(*cont.*)

criteria: presence of a chronic disease of the lungs, pulmonary vasculature, or respiratory gas exchange; plus presence of right ventricular enlargement or right heart failure

consider: right heart failure K77

K83 Heart valve disease NOS *I08, I34 to I37*

incl: chronic endocarditis; mitral valve prolapse; non-rheumatic aortic, mitral, pulmonary, tricuspid valve disorder

excl: rheumatic valve disease K71

criteria: absence of criteria for chronic rheumatic heart disease K71; plus evidence of valvular dysfunction by either characteristic heart murmur, or by imaging or echocardiographic evidence of abnormal valve

consider: hypertensive heart disease K87, cardiac murmurs unspecified K81

K84 Heart disease, other *I31, I42, I44 to I46, I51, O90.3*

incl: bundle branch block; cardiac arrest; cardiomegaly; disease of pericardium; cardiomyopathy; heart block; left bundle-branch block; other conduction disorders; other and unspecified heart disease

excl: cardiac arrhythmias K80

K85 Elevated blood pressure *R03.0*

incl: elevated blood pressure not meeting criteria for K86 and K87; transient or labile hypertension

K86 Hypertension, uncomplicated *I10*

incl: essential hypertension; hypertension NOS; idiopathic hypertension;

excl: hypertension with complications K87, in pregnancy W81

criteria: either two or more readings per encounter, taken at two or more encounters, with blood pressures that average over 95 mmHg diastolic or over 160 mmHg systolic in adult patients; or two or more readings at a single encounter with an average diastolic blood pressure of 120 mmHg or more; plus absence of evidence of secondary involvement of heart, kidney, eye, or brain

consider: elevated blood pressure K85

Notes 1: For children, consult appropriate paediatric blood pressure tables.

2: If secondary hypertension, code also the underlying cause.

K87 Hypertension, complicated *I11 to I13, I15, I67.4*

incl: malignant hypertension

excl: uncomplicated hypertension K86

criteria: blood pressure levels as defined in rubric K86, and abnormalities of the heart (enlargement, failure), kidney (albuminuria, azotemia), eye, or brain attributed to hypertension

K88 Postural hypotension *I95.0, I95.1, I95.8, I95.9*

incl: idiopathic, orthostatic hypotension
excl: due to drugs A85
criteria: signs or symptoms of cerebrovascular insufficiency (dizziness, syncope) on changing from the supine to the upright position; and a fall in mean blood pressure of 15 mmHg on two or more occasions when changing from the supine to the upright position
consider: low blood pressure K29

K89 Transient cerebral ischaemia *G45*

incl: basilar insufficiency; drop attacks; transient ischaemic attack (TIA); transient global amnesia
excl: cerebrovascular accident K90, migraine N89, carotid bruit K81
criteria: symptoms of transient (less than 24 hours) hypofunction of the brain, with sudden onset, presumed of vascular origin, without sequelae; and with exclusion of migraine, migraine equivalent, or epilepsy
consider: fainting, syncope A06
Note: double code with K91.

K90 Stroke/cerebrovascular accident *G46, I60 to I64*

incl: apoplexy; cerebral embolism; cerebral infarction; cerebral thrombosis; cerebrovascular accident (CVA), occlusion or stenosis; cerebral haemorrhage; subarachnoid haemorrhage
excl: transient cerebral ischaemia K89
criteria: signs and symptoms of a disturbance of cerebral function, presumed of vascular origin, lasting more than 24 hours or causing death, and within four weeks (28 days) of onset
Note: double code with K91.

K91 Cerebrovascular disease *I65, I66, I67.0 to I67.3, I67.5 to I67.9, I68, I69*

incl: cerebral aneurysm, sequelae of stroke
criteria: previous transient cerebral ischaemia or stroke; or investigation evidence of cerebrovascular disease

K92 Atherosclerosis/peripheral vascular disease *I70, I73, I74, R02*

incl: arteriosclerosis; arterial embolism, thrombosis or stenosis; atheroma; endarteritis; gangrene; intermittent claudication; limb ischaemia; Raynaud's syndrome; vasospasm
excl: mesenteric D99, ophthalmic F99, retinal F99, coronary K74 to K76, pulmonary K82, cerebral K89, K90, aneurysm K99, renal U99; and complicating abortion, pregnancy, childbirth W99, or puerperium W96

K93 Pulmonary embolism *I26*

incl: pulmonary (artery or vein) infarction, thromboembolism, thrombosis
excl: when complicating abortion, pregnancy, childbirth W99, or puerperium W96
criteria: sudden onset of dyspnoea or tachypnoea and either clinical or imaging evidence of pulmonary infarction, or ECG evidence of acute right ventricular strain
consider: dyspnoea R02

K94 Phlebitis and thrombophlebitis *I80 to I82, I87.0*

incl: superficial or deep vein thrombosis; phlebothrombosis; portal thrombosis
excl: cerebral thrombosis K90; thrombophlebitis in pregnancy W99 or in puerperium W96

K95 Varicose veins of leg *I83.1, I83.9, I87.2, I87.8*

incl: varicose eczema; venous insufficiency; venous stasis
excl: varicose ulcer S97
criteria: presence of dilated superficial veins in lower extremities; or demonstration of valve incompetence
consider: prominent veins K06

K96 Haemorrhoids *I84*

incl: internal haemorrhoids with or without complications; perianal haematoma; piles; residual haemorrhoidal skin tag; thrombosed external haemorrhoids; varicose veins of anus or rectum
criteria: visualization of varicosities of the venous plexus of the anus or canal; or tender, painful, blue-coloured localized swelling of acute onset, in the perianal area; or skin tags in the perianal area
consider: anal pain D04, rectal bleeding D16, anal lump D29

K99 Cardiovascular disease, other

I71, I72, I77, I78.0, I78.8, I78.9, I85, I86, I87.1, I87.9, I99, M30, M31, R57, T06.3, Z94.1, Z94.3

incl: aortic aneurysm; arteritis; arteriovenous fistula; lymphoedema; oesophageal varices; other aneurysm; polyarteritis nodosa; vasculitis; varicose veins of sites other than lower extremities
excl: lymphadenitis B71, cerebral aneurysm K91, gangrene K92

L Musculoskeletal

Component 1—Symptoms and complaints

L01 Neck symptom/complaint *M54.0, M54.2*

incl: pain attributed to cervical spine or musculoskeletal system
excl: headache, pain in head or face NOS N01

L02 Back symptom/complaint *M54.0, M54.6, M54.8, M54.9*

incl: backache NOS; thoracic back pain
excl: low back pain L03

L03 Low back symptom/complaint *M53.3, M54.0, M54.5*

incl: back pain (lumbar or sacroiliac); coccydynia; lumbago, lumbalgia
excl: thoracic back pain L02, sciatica L86

L04 Chest symptom/complaint *R29.8*

incl: chest pain attributed to musculoskeletal system
excl: chest pain NOS A11; precordial pain K01; painful respiration, pleuritic pain, pleurodynia R01

L05 Flank/axilla symptom/complaint *R29.8*

incl: loin pain
excl: kidney symptoms U14

(L06) deleted, included with L05

L07 Jaw symptom/complaint *M25.5, R29.8*

incl: temporomandibular joint symptoms
excl: teeth and gum symptoms/complaints D19

L08 Shoulder symptom/complaint *M25.4, M25.5, M25.6*

L09 Arm symptom/complaint *M79.6, R29.8*

excl: muscle pain, myalgia L18

L10 Elbow symptom/complaint *M25.4, M25.5, M25.6*

L11 Wrist symptom/complaint *M25.4, M25.5, M25.6*

L12 Hand and finger symptom/complaint *M25.4, M25.5, M25.6, R29.8*

L13 Hip symptom/complaint *M25.4, M25.5, M25.6, R29.4*

L14 Leg/thigh symptom/complaint *R25.2, R29.8*

incl: leg cramps
excl: muscle pain, myalgia L18; restless legs N04

L15 Knee symptom/complaint *M25.4, M25.5, M25.6*

L16 Ankle symptom/complaint *M25.4, M25.5, M25.6*

L17 Foot and toe symptom/complaint *M25.4, M25.5, M25.6, R29.8*

incl: metarsalgia

L18 Muscle pain
M60.1, M60.2, M60.8, M60.9, M77.4, M77.5, M79.0, M79.1, M79.6

incl: fibrositis, fibromyalgia, myalgia, rheumatism NOS
excl: pain in spine L01, L02, L03; leg cramps L14

L19 Muscle symptom/complaint NOS *M62.5, M62.6, M79.9*

incl: muscle stiffness; atrophy, wasting of muscle; muscle strain; weakness of muscle
excl: pain in spine L01, L02, L03; leg cramps L14; 'growing pains' in child L29; restless legs syndrome N04

L20 Joint symptom/complaint NOS *M25.4, M25.5, M25.6, M25.8, M25.9*

incl: arthralgia, effusion of joint, pain in joint, stiffness in joint, swelling of joint, weakness in joint
excl: specified joints, L07, L08, L10–13, L15–17

L26 Fear of cancer, musculoskeletal *Z71.1*

excl: if patient has cancer code the disease
criteria: concern on fear of cancer of musculoskeletal system in a patient without cancer or until the diagnosis is proven

L27 Fear of musculoskeletal disease, other *Z71.1*

excl: if patient has the disease code the disease
criteria: concern or fear of the disease in a patient without the disease or until the diagnosis is proven

L28 Limited function/disability musculoskeletal *Z73.6*

excl: falls A29; limping N29; walking difficulties N29; gait problems N29

Note: The COOP/WONCA Charts are suitable for documenting the patient's functional status (see Chapter 8).

L29 Musculoskeletal symptom/complaint, other *R29.3, R29.8*

incl: 'growing pains' in a child; other signs, symptoms and ill-defined conditions of musculoskeletal system

excl: clubbing of fingernails S22

Component 7—Diagnosis/disease

L70 Infections of musculoskeletal system

M00, M46.2 to M46.5, M60.0, M65.0, M65.1, M71.0, M71.1, M86

incl: infective tenosynovitis; osteomyelitis; pyogenic arthritis

excl: Reiter's disease L99, late effects of polio N70

criteria: infections localized to musculoskeletal system

L71 Malignant neoplasm, musculoskeletal *C40, C41, C49*

incl: fibrosarcoma; osteosarcoma

excl: secondary neoplasms (code to original site), benign/unspecified neoplasms L97

criteria: characteristic histological appearance

L72 Fracture: radius/ulna *S52*

incl: Colles' fracture

excl: pathological fracture L95, L99; non-union L99

criteria: imaging evidence of a fracture, or trauma plus visible or palpable deformity or crepitus involving the bone

consider: arm symptoms L09; other injury L81

L73 Fracture: tibia/fibula *S82.1 to S82.9*

incl: Pott's fracture

excl: fracture patella L76, pathological fracture L95, L99; non-union L99

criteria: imaging evidence of a fracture or trauma plus visible or palpable deformity or crepitus involving the bone

consider: leg symptoms L14, ankle symptoms L16, other injury L81

L74 Fracture: hand/foot bone *S62, S92*

incl: carpal, metacarpal bone; phalange of hand; phalange of foot; tarsal, metatarsal bone

(*cont.*)

excl: pathological fracture L95, L99; non-union L99
criteria: imaging evidence of a fracture, or trauma plus visible or palpable deformity or crepitus involving the bone
consider: arm L09 or leg L14 symptoms; other injury L81

L75 Fracture: femur *S72*

incl: fracture neck of femur
excl: pathological fractures L95, L99; non-union L99
criteria: imaging evidence of a fracture, or trauma plus visible or palpable deformity or crepitus involving the bone
consider: leg symptoms L14; other injury L81

L76 Fracture: other

S02.2 to S02.4, S02.6 to S02.9, S12, S22, S32, S42, S82.0, T08, T10, T12, T14.2

incl: all fractures not specified in L72 to L75, N80
excl: pathological fracture L95, L99; non-union L99; skull fracture N80
criteria: imaging evidence of a fracture, or trauma plus visible or palpable displacement of the bone surface
consider: symptoms in Component 1

L77 Sprains and strains of ankle *S93.4*

criteria: stretch injury of the affected part plus pain aggravated by stretching or tensing the affected structure
consider: ankle symptoms L16

L78 Sprains and strains of knee *S83.4, S83.6*

excl: acute damage of meniscus/internal ligament of knee L96
criteria: stretch injury of the affected part plus pain aggravated by stretching or tensing the affected structure

L79 Sprains and strains of joints NOS

S03.4, S03.5, S23.3 to S23.5, S33.6, S43.4 to S43.7, S53.2 to S53.4, S63.3 to S63.7, S73.1, S93.2, S93.5, S93.6, T14.3

incl: all sprains and strains of joints and ligaments except ankle, knee, and spine
excl: ankle L77; knee L78; cervical spine, whiplash injury L83; back strain L84
criteria: stretch injury of the affected part plus pain aggravated by stretching or tensing the affected structure
consider: symptoms in Component 1

L80 Dislocation and subluxation

M22.0, M22.1, S03.0, S03.3, S13.0 to S13.3, S23.1, S23.2, S33.1 to S33.3, S43.0 to S43.3, S53.0, S53.1, S63.0 to S63.2, S73.0, S83.0, S83.1, S93.0, S93.1, S93.3, T14.3

incl: all dislocations and subluxations including spine
criteria: trauma to the joint plus either imaging evidence of a dislocation or subluxation, or palpable or visible dislocation deformity
consider: symptoms in Component 1
Note: code fracture dislocations to the fracture.

L81 Injury musculoskeletal NOS

M79.5, S09.1, S16 to S19, S23.0, S28, S29, S33.0, S33.4, S39, S46 to S49, S56 to S59, S66 to S69, S76 to S79, S86 to S89, S96 to S99, T06.4, T11, T13, T14.6, T14.7

incl: deep foreign body; haemarthrosis; traumatic amputation
excl: internal injury of chest, abdomen, and pelvis A81; multiple trauma A81; late effects of trauma: deformities, disabilities, scarring A82; injuries of teeth D80, and eardrum H77; non-union L99; traumatic arthropathy L91; mal-union of fracture L99; head injury, concussion, intracranial injury, skull fracture N80; laceration and other injury to nerve N81; insect bite and sting S12; animal bite S13; bruise, contusion S16; laceration, open wound S18

L82 Congenital anomaly musculoskeletal *Q65 to Q79*

incl: bow leg; clubfoot (talipes); congenital dislocation of hip; genu recurvatum; malformations of skull and face; other congenital deformities of the foot
excl: scoliosis L85; pes planus (acquired) L98; spina bifida N85

L83 Neck syndrome

M43.3, M43.4, M43.6, M47.8, M47.9, M48, M50, M53.0, M53.1, S13.4 to S13.6

incl: syndromes with and without radiation of pain: cervical disc lesion; cervicobrachial syndrome; osteoarthritis; radicular syndrome of upper limbs; spondylosis; sprains and strains; torticollis; whiplash; cervicogenic headache

L84 Back syndrome without radiating pain

M43.0, M43.1, M43.5, M46.0, M46.1, M46.8, M46.9, M47.0, M47.8, M47.9, M48, M51, M53.2 to M53.9, S33.5, S33.7

incl: back strain; collapsed vertebra NOS; facet joint degeneration; osteoarthrosis, osteoarthritis of spine; spondylosis; spondylolisthesis
excl: coccydynia L03; syndromes related to the neck L83; back pain with radiation L86; sciatica L86; psychogenic backache P75
criteria: back pain without radiation plus limitation of movement confirmed at medical examination
consider: symptoms/complaints back L02; low back L03

L85 Acquired deformity of spine *M40, M41, M43.8, M43.9*

incl: kyphoscoliosis; kyphosis; lordosis; scoliosis
excl: congenital deformities L82; ankylosing spondylitis L88; spondylolisthesis L84

L86 Back syndrome with radiating pain *M47.1, M47.2, M51, M54.3, M54.4*

incl: disc prolapse or degeneration; sciatica
excl: cervical disc lesion L83; spondylolisthesis L84; recent back strain L84
criteria: pain in the lumbar or thoracic region of the spine, accompanied by pain radiating to, or a neurological deficit of, an appropriate area; or sciatica, pain radiating down the back of the leg, aggravated by coughing, movement, or posture; or demonstration of a prolapsed lumbar or thoracic disc by appropriate imaging technique, or at surgery
Note: exclude referred pain which is diffuse.
consider: back pain L02, L03

L87 Bursitis/tendonitis/synovitis NOS

M65.2 to M65.4, M65.8, M65.9, M67.3, M67.4, M70, M71.2 to M71.9, M72, M76, M77.0, M77.2, M77.3, M77.8, M77.9

incl: bone spurs, calcified tendon, Dupuytren's contracture, fasciitis, ganglion, synovial cysts, tenosynovitis, trigger finger
excl: of shoulder L92; tennis elbow, lateral epicondylitis L93

L88 Rheumatoid arthritis *M05, M06, M08, M45*

incl: allied conditions: ankylosing spondylitis; juvenile arthritis
excl: psoriatic arthropathy S91

L89 Osteoarthrosis of hip *M16*

incl: osteoarthritis; secondary to dysplasia, trauma
criteria: either characteristic imaging appearance; or joint disorder of at least 3 months' duration, with no constitutional symptoms and three or more of the following: intermittent swelling; crepitation; stiffness or limitation of movement; normal ESR, rheumatoid tests, and uric acid; over 40 years of age
consider: joint symptoms L20; arthritis NOS L91

L90 Osteoarthrosis of knee *M17*

incl: osteoarthritis, secondary to trauma
criteria: either characteristic imaging appearance, or joint disorder of at least 3 months' duration, with no constitutional symptoms and three or more of the following: intermittent swelling; crepitation; stiffness or limitation of movement; normal ESR, rheumatoid tests, and uric acid; over 40 years of age
consider: joint symptoms L20; arthritis NOS L91

L91 Osteoarthrosis, other *M13, M15, M18, M19*

incl: arthritis NOS; Heberden's nodes; osteoarthritis; traumatic arthropathy
excl: of neck L83; of spine L84; of hip L89; of knee L90; of shoulder L92
criteria: characteristic imaging appearance; or Herberden's nodes; or joint disorder of at least 3 months' duration, with no constitutional symptoms and three or more of the following: intermittent swelling; crepitation; stiffness or limitation of movement; normal ESR, rheumatoid tests, and uric acid over 40 years of age
consider: joint symptoms L20; arthritis NOS L91

L92 Shoulder syndrome *M75*

incl: bursitis of shoulder; frozen shoulder; osteoarthrosis; rotator cuff syndrome; synovitis of shoulder; tendinitis around shoulder
criteria: shoulder pain with limitation of movement or local tenderness or crepitus; or periarticular calcification on imaging

L93 Tennis elbow *M77.1*

incl: lateral epicondylitis
excl: other tendonitis L87

L94 Osteochondrosis *M42, M91, M92, M93*

incl: Legg–Calve–Perthes disease; Osgood/Schlatter disease; osteochondritis dissecans; Scheuermann's disease; slipped femoral epiphysis

L95 Osteoporosis *M80, M81*

incl: pathological fractures due to osteoporosis
criteria: characteristic imaging appearance

L96 Acute internal damage knee *S83.2, S83.3, S83.5, S83.7*

incl: acute damage to meniscus or cruciate ligaments
excl: acute damage to collateral ligaments L78, dislocation of patella L80, chronic internal damage to knee L99
criteria: an initial injury which occurred no longer than one month previously and demonstration of ligament or meniscus tear by surgery, arthroscopy, or imaging; or by locking or giving way, pain, and swelling of knee
consider: knee symptoms L15; sprain of knee L78

L97 Neoplasm, benign/uncertain musculoskeletal *D16, D21, D48.0, D48.1*

incl: benign and unspecified neoplasms of musculoskeletal system
excl: malignant neoplasms L71

L98 Acquired deformity of limb *M20, M21*

incl: bunion; genu valgum-varum; hallux valgus/varus; mallet finger; pes planus (flatfoot)
excl: congenital deformities and anomalies A90, L82

L99 Musculoskeletal disease, other

M02, M12, M22.2 to M22.9, M23, M24, M25.0 to M25.3, M25.7 to M25.9, M32 to M35, M43.2, M54.1, M61, M62.0 to M62.4, M62.8, M62.9, M66, M67.0 to M67.2, M67.8, M67.9, M79.3, M79.4, M79.8, M84, M85, M87 to M89, M94, M95, M99, T79.6, Z89, Z94.6, Z98.1

incl: arthrodesis; chronic internal derangement of knee; contractures; costochondritis; dermatomyositis; disorders of patella; malunion or non-union of fracture; myositis; Paget's disease of bone; panniculitis; pathological fracture NOS; polymyalgia rheumatica; psoriatic arthritis (code also S91); Reiter's disease; scleroderma; Sjogren's syndrome; spontaneous rupture tendon; systemic lupus erythematosis
excl: hyperuricemia A91; pathological fracture due to osteoporosis L95; post-polio paralysis N70; post-stroke paralysis N81; gout T92; pseudogout and crystal arthropathies T99; osteomalacia T99

N Neurological

Component 1—Symptoms and complaints

N01 Headache *G44.3, G44.8, R51*

incl: pain in head, post-traumatic headache
excl: face pain N03; migraine N89; cluster headache N90; tension headache N95; atypical facial neuralgia N99; sinus pain R09; post-herpetic pain S70

(N02) transferred to N95

N03 Pain, face *G50.1, R51*

excl: toothache D19, headache N01, migraine N89, trigeminal neuralgia N92, sinus pain R09, post-herpetic pain S70

N04 Restless legs *G25.8*

excl: leg cramps L14, intermittent clandication K92

N05 Tingling fingers/feet/toes *R20.2*

incl: burning sensation, prickly feeling
excl: pain, tenderness of skin S01

N06 Sensation disturbances, other *R20.0, R20.l, R20.3, R20.8*

incl: anaesthesia, numbness, paraesthesia
excl: tingling N05; pain, tenderness of skin S01

N07 Convulsions/seizures *R56*

incl: febrile convulsions, fits
excl: fainting A06, transient ischaemic attack K89

N08 Abnormal involuntary movements

G25.0 to G25.5, G25.8, G25.9, R25.0, R25.1, R25.3, R25.8

incl: dystonia, jerking, myoclonus, shaking, spasm, tetany, tremor, twitching
excl: chorea K71; cramps L12, L14, L17, L18; restless legs N04; convulsions N07; tic douloureux N92; organic tics N99; psychogenic tic P10

N16 Disturbance of smell/taste *R43*

incl: anosmia
excl: halitosis D20

N17 Vertigo/dizziness *R42*

incl: giddiness, feeling faint, lightheaded, loss of balance, woozy
excl: syncope A06; blackout A06; motion sickness A88; specific vertiginous syndromes H82

N18 Paralysis/weakness *G98*

incl: paresis
excl: general weakness A04

N19 Speech disorder *R47*

incl: aphasia, dysphasia, dysarthria, slurred speech
excl: stammering, stuttering P10; speech delay P22; hoarseness R23

N26 Fear of cancer of neurological system *Z71.1*

excl: if patient has the disease code the disease
criteria: fear of cancer in a patient without the disease or until the diagnosis is proven

N27 Fear of neurological disease, other *Z71.1*

excl: if patient has the disease code the disease
criteria: fear of other neurological disease in a patient without the disease or until the diagnosis is proven

N28 Limited function/disability neurological *Z73.6*

incl: disability due to neurological diseases and disorders
Note: The COOP/WONCA Charts are suitable for documenting the patient's functional status (see Chapter 8).

N29 Neurological symptom/complaint, other
M79.2, R26, R27, R29.0 to R29.2, R29.8

incl: gait abnormality; limping; ataxia; meningism

Component 7—Diagnosis/diseases

N70 Poliomyelitis *A80, A85.0, B91*

incl: late effects, post-polio syndrome, other enterovirus infections

N71 Meningitis/encephalitis
A32.1, A39.0, A83, A84, A85.1, A85.2, A85.8, A86, A87, B00.3, B00.4, B37.5, B58.2, B94.1, G00, G03 to G04

excl: myalgic encephalomyelitis A04
criteria: an acute febrile illness with abnormal findings in the cerebro-spinal fluid.
consider: fever A03, meningism N29

N72 Tetanus *A33 to A35*

excl: tetany N08
criteria: rigidity, hypertonic contractions or tetanic spasticity and a history of preceding injury

N73 Neurological infection, other *A81, A88.8, A89, G06, G08 to G09*

incl: cerebral abscess, slow virus infections
excl: poliomyelitis N70; meningitis/encephalitis N71; acute polyneuritis N94

N74 Malignant neoplasm nervous system *C47, C70 to C72*

criteria: characteristic histological appearance
consider: unspecified neoplasm N76

N75 Benign neoplasm nervous system *D32, D33, D36.1*

incl: acoustic neuroma, meningioma

N76 Neoplasm nervous system, uncertain nature *D42, D43, D48.2*

excl: neurofibromatosis A90

N79 Concussion *S06.0*

incl: late effects
excl: psychological effects P02
criteria: trauma to the head with a temporary loss of consciousness, often accompanied by amnesia, without serious cerebral contusion
consider: other head injury N80

N80 Head injury, other *S02.0, S02.1, S06.1 to S06.9, S07, S08, S09.0, S09.7, S09.8*

incl: cerebral injury with or without skull fracture; cerebral contusion; traumatic intracerebral haemorrhage; extradural haematoma; subdural haematoma
excl: concussion N79
criteria: trauma to the head, complicated by cerebral damage

N81 Injury nervous system, other

S04, S09.9, S14, S24, S34, S44, S54, S64, S74, S84, S94, T06.0 to T06.2, T14.4

incl: nerve injury, spinal cord injury

N85 Congenital anomaly neurological *Q00 to Q07*

incl: hydrocephalus, spina bifida

N86 Multiple sclerosis *G35*

incl: disseminated sclerosis
criteria: exacerbations and remissions of multiple neurological manifestation with deficits/derangements disseminated in both time and site (any combination of neurological signs and symptoms is possible)
consider: other neurological symptoms N29

N87 Parkinsonism *G20, G21*

incl: drug induced; paralysis agitans; Parkinson's disease
criteria: poverty and slowness of voluntary movements, resting tremor improving with active purposeful movement, and muscular rigidity
consider: abnormal involuntary movements N08, disorder of speech N19

N88 Epilepsy *G40, G41*

incl: all types of epilepsy, focal seizures, generalized seizures, grand mal, petit mal, status epilepticus
criteria: recurrent episodes of sudden altered consciousness, with or without tonic or clonic movements/seizures, plus either eyewitness account of the attack, or characteristic abnormality of electroencephalogram (EEG)
consider: convulsions N07, other neurological symptoms N29

N89 Migraine *G43, G44.1*

incl: vascular headache with and without aura
excl: cervicogenic headache L83, cluster headache N90, tension headache N95
criteria: recurrent episodes of headache with three or more of unilateral headache, nausea or vomiting, aura, other neurological symptoms, family history of migraine
consider: headache N01

N90 Cluster headache *G44.0*

excl: migraine N89
criteria: attacks of severe, often excruciating unilateral pain peri-orbitally and/or temporally, occurring up to 8 times a day, sometimes associated with conjunctival injection, lacrimation, nasal congestion, rhinorrhoea, sweating, miosis, ptosis, or eyelid oedema. Attacks occur in cluster periods lasting for weeks or months separated by remissions lasting months or years.

N91 Facial paralysis/Bell's palsy *G51*

criteria: acute onset of unilateral paralysis of muscles of facial expression without sensory loss
consider: paralysis, weakness N18

N92 Trigeminal neuralgia *G50.0, G50.8, G50.9*

incl: tic douloureux
excl: post-herpetic neuralgia S70
criteria: unilateral paroxysms of burning facial pain aggravated by touching trigger-points, blowing nose or yawning, without sensory or motor paralysis
consider: neuralgia NOS N99

N93 Carpal tunnel syndrome *G56.0*

criteria: loss or impairment of superficial sensation affecting the thumb, index and middle finger, that may or may not split the ring finger. Dysesthesia and pain worsen usually during the night, and may radiate to the forearm
consider: sensation disturbances N06

N94 Peripheral neuritis/neuropathy
G51, G54, G56.1 to G56.4, G56.8, G56.9, G57, G60 to G62, G64

incl: acute infective polyneuropathy, diabetic neuropathy (double code with T89 or T90), Guillain-Barre syndrome, nerve lesion, neuropathy, phantom limb
excl: post-herpetic S70
criteria: sensory, reflex and motor changes confined to the territory of individual nerves sometimes without apparent cause, sometimes secondary to a specific disease, e.g. diabetes

N95 Tension headache *G44.2*

excl: migraine N89; cluster headache N90
criteria: pressing, generalized headache associated with stress and muscle tension with or without increased tenderness of pericranial muscles
consider: headache N01

N99 Neurological disease, other

E51.2, G10 to G12, G23, G24, G31.0, G31.1, G31.8, G31.9, G36, G37, G52, G70 to G72, G80 to G83, G90 to G92, G93.0 to G93.2, G93.4 to G93.9, G95 to G96, G98, G99, M79.2, Z98.2

incl: cerebral palsy, motor neurone disease, myasthenia gravis, neuralgia NOS
excl: sleep apnoea P06

P Psychological

Component 1—Symptoms and complaints

P01 Feeling anxious/nervous/tense *F41.9, R45.0*

incl: anxiety NOS, feeling frightened
excl: anxiety disorder P74
criteria: feelings reported by the patient as an emotional or psychosocial experience not attributed to the presence of a mental disorder. A gradual transition exists from feelings that are unwelcome—but quite normal—to feelings that are so troublesome to the patient that professional help is sought

P02 Acute stress reaction *F43.0, F43.2, F43.8, F43.9, F50.4*

incl: adjustment disorder, culture shock, feeling stressed, grief, homesick, immediate post-traumatic stress, shock (psychic)
excl: feeling depressed P03, depressive disorder P76, post-traumatic stress disorder P82
criteria: a reaction to a stressful life event or significant life change requiring a major adjustment, either as an expected response to the event or as a maladaptive response interfering with daily coping and resulting in impaired social functioning, with recovery within a limited period of time

P03 Feeling depressed *R45.2, R45.3*

incl: feeling inadequate, lonely, unhappy, worried
excl: depressive disorder P76, low self-esteem P28
criteria: feelings reported by the patient as an emotional or psychological experience not attributed to the presence of a mental disorder. A gradual transition exists from feelings that are unwelcome—but quite normal—to feelings that are so troublesome to the patient that professional help is sought

P04 Feeling/behaving irritable/angry *R45.1, R45.4, R45.5, R45.6*

incl: agitation NOS, restlessness NOS

excl: overactive child P81, irritability in family member Z13

criteria: feelings reported by the patient as an emotional or psychological experience not attributed to the presence of a mental disorder, or behaviour indicating irritability or anger. A gradual transition exists from feelings or behaviour that are unwelcome—but quite normal—to those that are so troublesome that professional help is sought

P05 Senility, feeling/behaving old *R54*

incl: concern with aging, senescence

criteria: feelings reported by the patient as an emotional or psychological experience not attributed to the presence of a mental disorder. A gradual transition exists from feelings that are unwelcome—but quite normal—to feelings that are so troublesome to the patient that professional help is sought

P06 Sleep disturbance *F51, G47*

incl: insomnia, nightmares, sleep apnoea, sleepwalking, somnolence

excl: jetlag A88

criteria: sleep disturbance as a diagnosis requires that the sleeping problem forms a major complaint, which, according to both patient and doctor, is not caused by another disorder but is a condition in its own right. Insomnia requires a quantitative or qualitative deficiency of sleep which is unsatisfactory in the patient's opinion, over a considerable period of time. In hypersomnia excessive daytime sleepiness and sleep attacks exist which limit the patient's performance

P07 Sexual desire reduced *F52.0*

incl: frigidity, loss of libido

excl: impotence P08, loss of sexual fulfilment P08, concern with sexual preference P09

criteria: sexual problems with regard to desire not caused by any organic disorder or disease, but a reflection of the inability of a patient to participate in the sexual relationship s/he wants because of lack of desire, failure of genital response or function

P08 Sexual fulfilment reduced *F52.1 to F52.9*

incl: non-organic impotence or dyspareunia, premature ejaculation, vaginismus of psychogenic origin

excl: sexual problems with desire P07, concern with sexual preference P09, organic impotence and sexual problems Y07, vaginismus NOS X04

criteria: sexual problems with regard to fulfilment not caused by any organic disorder or disease, but a reflection of the inability of a patient to participate in

(*cont.*)

the sexual relationship s/he wants because of failure of genital response or function, or problems with sexual development

P09 Sexual preference concern *F64 to F66*

excl: sexual problems with desire P07 or fulfilment P08

criteria: sexual problems with regard to preference not caused by any organic disorder or disease, but a reflection of the inability of a patient to participate in the sexual relationship s/he wants because of problems with sexual identity, preference, or orientation

P10 Stammering, stuttering, tics *F95, F98.4, F98.5, F98.6*

excl: tic douloureux N92

criteria: stammering and stuttering: disorder of speech characterized by frequent repetitions or prolongations of sounds, or by frequent hesitations on pauses disrupting speech

P11 Eating problems in children *F98.2, F98.3*

incl: feeding problem, problem with eating behaviour of children

excl: anorexia nervosa P86, eating problem in adults T05

Note: Problems with behaviour of children are particularly difficult to classify, which is illustrated by the fact that they are distributed over four chapters of ICPC. Whether or not parents present these problems will reflect their ideas about the gradual differences between normal—though maybe annoying—behaviour, and behaviour that is considered worrying or 'pathological'.

P12 Bedwetting, enuresis *F98.0*

excl: due to organic disorders U04

criteria: involuntary voiding of urine by day or by night not determined to be related to any organic disorder

Note: Problems with behaviour of children are particularly difficult to classify, which is illustrated by the fact that they are distributed over four chapters of ICPC. Whether or not parents present these problems will reflect their ideas about the gradual differences between normal—though maybe annoying—behaviour, and behaviour that is considered worrying or 'pathological'.

P13 Encopresis/bowel training problem *F98.1*

criteria: encopresis requires repeated passage of usually well formed faeces in inappropriate places, considered abnormal in relation to age, and not caused by constipation, sphincter control disorder, or other disease

P15 Chronic alcohol abuse *F10.1 to F10.9, G31.2*

incl: alcoholism, alcohol brain syndrome, alcohol psychosis, delirium tremens

(cont.)

criteria: a disorder due to the use of alcohol resulting in one or more of harmful use with clinically important damage to health, dependence syndrome, withdrawal state, or psychotic disorder

P16 Acute alcohol abuse *F10.0*

incl: drunk
criteria: a disorder due to the use of alcohol resulting in acute intoxication, with or without a background of chronic abuse

P17 Tobacco abuse *F17*

incl: smoking problem
criteria: a disorder due to the use of tobacco resulting in one or more of acute intoxication, harmful use with clinically important damage to health, dependence syndrome, or withdrawal state
consider: risk factor NOS A23

Note: Substance abuse problems should take into account the considerable differences between countries and cultures. An alcohol dependent or heroin addicted patient needs medical attention, but the definitions of 'tobacco abuse' are controversial. In these cases doctors can decide to label the episode as 'tobacco abuse' without the patient's agreement, and consequently also without their willingness to agree to any medical intervention.

P18 Medication abuse *F13, F19, F55*

incl: any prescribed drugs

Note: Substance abuse problems should take into account the considerable differences between countries and cultures. An alcohol dependent or heroin addicted patient needs medical attention, but some patients request and use tranquillizers, sleeping tablets, anorectics, or laxatives inappropriately and for too long. In these cases doctors can decide to label the episode as 'medicine abuse' without the patient's agreement, and consequently also without their willingness to agree to any medical intervention.

P19 Drug abuse *F11 to F16, F18, F19*

incl: addiction to drugs, drug withdrawal
criteria: a disorder due to the use of a dependence producing psychoactive substance, resulting in one or more of the following conditions:
—acute intoxication
—harmful use with clinically important damage to health
—dependence syndrome
—withdrawal state
—psychotic disorder

Note: Substance abuse problems should take into account the considerable differences between countries and cultures. An alcohol dependent or heroin addicted patient needs medical attention, but the definitions of 'use of hashish' are controversial. Some

(*cont.*)

patients request and use tranquillizers, sleeping tablets, anorectics, or laxatives inappropriately and for too long. In these cases doctors can decide to label the episode as 'drug abuse' without the patient's agreement, and consequently also without their willingness to agree to any medical intervention.

P20 Memory disturbance *R41*

incl: amnesia; disturbance of concentration; disorientation

(P21) transferred to P81

P22 Child behaviour symptom/complaint

F91 to F94, F98.8, F98.9, R62.0

incl: delayed milestones; jealousy; overactive child; speech delay; temper tantrum

excl: adolescent P23; adult P80; concern about physical development T10; growth delay T10

P23 Adolescent behaviour symptom/complaint

F91, F92, F94, F98.8, F98.9

incl: delinquency

excl: child P22, P81; adult P80

P24 Specific learning problem *F80 to F83, R48*

incl: dyslexia

excl: attention deficit disorder P81, mental retardation P85

criteria: specific speech, language, and learning problems with onset in childhood, together with an impairment of functions related to biological maturation of the central nervous system, and a steady course over time without spontaneous remissions or relapses, although the deficit may diminish as the child grows older

P25 Phase of life problems in adults *Z60.0*

incl: empty nest syndrome, mid-life crisis, retirement problem

excl: senility, feeling/behaving old P05; menopause X11

P27 Fear of mental disorder *Z71.1*

incl: concern about mental disease, fear of attempting suicide

excl: if patient has the disease code the disease

criteria: concern or fear of mental disease in a patient without the disease or until the diagnosis is proven

P28 Limited function/disability psychological *Z73.6*

incl: limited function/disability as a result of psychological disorders, low self-esteem

Note: The COOP/WONCA Charts are suitable for documenting the patient's functional status (see Chapter 8).

P29 Psychological symptom/complaint, other
F50.8, F50.9, F98.8, F98.9, R44, R45.7, R45.8, R46, Z64.2, Z64.3, Z73.0, Z73.1

incl: delusions, eating disorders NOS, hallucinations, multiple psychological symptoms/complaints, poor hygiene, strange behaviour, suspiciousness

excl: tension headache N95

Component 7—Diagnosis/diseases

Note: A mental disorder is a clinically significant psychological syndrome or pattern, with or without an association with stressors (such as disability, increased risk, or an important loss), which cannot be considered an expected response to a particular event, but rather a manifestation of a behavioural, psychological or biological dysfunction.

P70 Dementia *F01 to F03, G30*

incl: Alzheimer's disease, senile dementia

criteria: a syndrome due to a disease of the brain, usually of a chronic and/or progressive nature, with clinically significant disturbance of multiple higher cortical functions (memory, thinking, orientation, comprehension), together with intact consciousness

consider: senility P05, other psychological symptoms P29

P71 Organic psychosis, other *F04 to F07, F09*

incl: delirium

excl: alcoholic cause P15, unspecified psychosis P98

criteria: organic psychiatric disorders as a diagnosis require psychological syndromes, patterns or behaviour due to organic disease

P72 Schizophrenia *F20 to F22, F24, F25, F28, F29*

incl: all types of schizophrenia, paranoia

excl: acute and transient psychosis P98

criteria: fundamental and characteristic distortions of thinking, perceptions, and affects that are inappropriate or blunted (e.g. thought-echo, -insertion, -withdrawal, delusional perceptions, hallucinatory voices, delusions of control), together with a clear consciousness and unaffected intellectual capacity.

consider: psychosis NOS P98

P73 Affective psychosis *F30, F31, F34.0*

incl: bipolar disorder, hypomania, mania, manic depression
excl: depression P76
criteria: a fundamental disturbance in affect and mood, alternately being elated and depressed (with/without associated anxiety). In manic disorder mood, energy and activity are simultaneously elevated. In bipolar disease, at least two periods of disturbed mood, shifting from elevated to lowered are observed
consider: psychosis NOS P98

P74 Anxiety disorder/anxiety state *F41.0, F41.1, F41.3 to F41.9*

incl: anxiety neurosis, panic disorder
excl: anxiety with depression P76, anxiety NOS P01
criteria: clinically significant anxiety that is not restricted to any particular environmental situation. It manifests as a panic disorder (recurrent attacks of severe anxiety not restricted to any particular situation, with or without physical symptoms) or as a disorder in which generalized and persistent anxiety, not related to any particular situation, occurs with variable physical symptoms
consider: feeling anxious, nervous, tense P01

P75 Somatization disorder *F44, F45*

incl: conversion disorder, hypochondriacal disorder, hysteria, pseudocyesis
criteria: somatization disorder is characterized by a preoccupation with and repeated presentations of physical symptoms and complaints together with persistent requests for medical investigations in spite of repeated negative findings and reassurances by doctors. For this diagnosis, the presentation of multiple, recurrent, and frequently changing physical symptoms presented to the general/family practitioner over a period of at least one year is required. Hypochondriacal disorder requires a persistent preoccupation with either the physical appearance or with the possibility of having a serious disease, together with persistent somatic complaints over a period of at least one year, in spite of repeated negative findings and reassurances by doctors

Notes: Somatization is the repeated presentation of physical symptoms and complaints suggesting physical disorders for which no demonstrable organic findings or physiological mechanisms are responsible, and for which there is positive evidence that they are linked to psychological factors while the patient does not experience a sense of controlling the production of these symptoms in dealing with the psychological factors. Physical symptoms and complaints including pain which are presented as if they were due to a physical disorder of a system/organ under autonomic nervous control, or which consist of persistent, severe/distressing pain which cannot be explained by a physiological process/disorder, are coded with a symptom/complaint diagnosis representing the physical aspect, and—if possible—with a code representing the emotional or psychosocial problem with which it is associated.

The definition of somatization disorder in ICD-10 (a minimum of two years) is too stringent for use in general/family practice.

P76 Depressive disorder *F32, F33, F34.1, F34.8, F34.9, F38, F39, F41.2, F53.0*

incl: depressive neurosis; depressive psychosis; mixed anxiety and depression; reactive depression; puerperal/post-natal depression

excl: acute stress reaction P02

criteria: fundamental disturbance in affect and mood towards depression. Mood, energy and activity are simultaneously lowered, together with an impaired capacity for enjoyment, interest and concentration. Sleep and appetite are usually disturbed and self-esteem and confidence are decreased

consider: feeling depressed P03

P77 Suicide/suicide attempt *F99*

incl: suicide gesture; successful attempt (double code with A96)

excl: fear of committing suicide P27

P78 Neuraesthenia, surmenage *F48.0*

criteria: increased fatiguability with unpleasant associations, difficulties in concentration and a persistent decrease in performance and coping efficiency; the feeling of physical weakness and exhaustion after mental effort or after a minimal physical effort is often accompanied by muscular pain and an inability to relax

consider: fatigue A04, post-viral fatigue syndrome A04, chronic fatigue syndrome A04

P79 Phobia, compulsive disorder *F40, F42, F48.1, F48.8, F48.9*

incl: phobias, obsessive-compulsive disorder

criteria: phobic anxiety disorder requires outspoken anxiety, evoked only in well defined situations that are not generally considered dangerous: the patient tries to avoid these situations, or endures them with dread.
Obsessive compulsive disorder requires distressing and recurrent obsessional thoughts or acts recognized by the patient as his/her own; compulsive stereotyped behaviours are repeated again and again, intended to prevent some objective unlikely event and recognized by the patient as pointless and ineffective

P80 Personality disorder *F60 to F63, F68, F69, Z76.5*

incl: psychopath, compensation neurosis, Munchhausen's syndrome, adult behaviour disorder

criteria: persistent and clinically important conditions and behaviour patterns in an individual's lifestyle and mode of relating to him/herself and others, reflecting significant or extreme deviations from the way an average individual in a given culture perceives, feels and behaves. This pattern is deeply ingrained and longlasting

P81 Hyperkinetic disorder *F90*

incl: attention deficit disorder (ADD), hyperactivity
excl: adolescent onset P23, learning disorders P24
criteria: early onset of a lack of persistence in activities requiring cognitive involvement, with a tendency to move from one activity to another without completing any one, with disorganized and ill-regulated behaviour, and excessive activity
consider: overactive child P22

P82 Post-traumatic stress disorder *F43.1*

incl: persistent adjustment disorder
criteria: a stressful event followed by a major state of distress and disturbance, with a delayed or protracted reaction, flashbacks, nightmares, emotional blunting, and anhedonia interfering with social functioning and performance, and including depressed mood, anxiety, worry, and feeling unable to cope, persistent over time
consider: acute stress reaction P02, feeling anxious P01, feeling depressed P03

P85 Mental retardation *F70 to F79*

excl: due to congenital anomalies A90
criteria: arrested or incomplete development of the mind with impairment of skills during the developmental period, and a low overall level of intelligence, with or without impairment of behaviour.

P86 Anorexia nervosa, bulimia *F50.0 to F50.2, F50.3, F50.4*

criteria: Anorexia nervosa: deliberate weight loss induced and sustained by the patient, associated with an intensive and overvalued dread of fatness and flabbiness of body contures.
Bulimia: repeated bouts of overeating and an excessive preoccupation with body weight, leading to a pattern of overeating followed by induced vomiting or use of purgatives
consider: eating disorders, food refusal P11, P29; feeding problem T04, T05

P98 Psychosis NOS, other *F23, F53.1*

incl: acute and transient psychosis, reactive psychosis, chronic brain syndrome, puerperal psychosis

P99 Psychological disorders, other *F53.8, F53.9, F54, F59, F84, F88, F89, F99*

incl: autism; mental illness unspecified; neurosis NOS

R Respiratory

Component 1—Symptoms and complaints

R01 Pain, respiratory system *R07.1*

incl: painful respiration, pleuritic pain, pleurodynia
excl: chest pain A11, musculo-skeletal chest pain L04, nose pain R08, sinus pain R09, sore throat R21, pleurisy R82, chest tightness R29

R02 Shortness of breath, dyspnoea *R06.0*

incl: orthopnoea
excl: wheezing R03, stridor R04, hyperventilation R98

R03 Wheezing *R06.2*

incl: inspiratory wheeze or rhonchi
excl: dyspnoea R02, stridor R04, hyperventilation R98

R04 Breathing problems, other *R06.1, R06.3, R06.5, R06.8*

incl: abnormal breathing, apnoea, holding breath, respiratory distress, snoring, stridor, tachypnoea
excl: sleep apnoea P06, respiratory pain R01, dyspnoea R02, wheezing R03, cough R05, hyperventilation syndrome R98

R05 Cough *R05*

incl: cough (dry or moist)
excl: sputum or phlegm R25

R06 Nose bleed/epistaxis *R04.0*

R07 Sneezing/nasal congestion *R06.7*

incl: blocked nose, rhinorrhoea, running nose

R08 Nose symptom/complaint, other *J34.8*

incl: pain in nose, post-nasal drip, prominent nose, red nose
excl: anosmia N16, epistaxis R06, blocked nose R07, sneezing R07, complaints of sinuses R09, rhinophyma S99

R09 Sinus symptom/complaint, other *J34.8*

incl: blocked sinuses, congested sinuses, pain or pressure in sinuses
excl: headache N01, face pain N03, nasal congestion R07

R21 Throat symptom/complaint *R07.0, R09.8*

incl: dry throat, inflamed throat, large tonsils, lump in throat, red throat, sore throat, tonsillar pain
excl: voice symptoms R23, tonsillar hypertrophy R90

(R22) deleted, amalgamated with R21

R23 Voice symptom/complaint *R49*

incl: absence of voice, aphonia, hoarseness
excl: neurological disorders of speech N19, stammering/stuttering/tics P10, sore throat R20

R24 Haemoptysis *R04.2*

incl: coughing blood

R25 Sputum/phlegm abnormal *R09.3*

excl: cough with sputum R05, haemoptysis R24

R26 Fear of cancer of respiratory system *Z71.1*

excl: if patient has the disease code the disease
criteria: fear of cancer of respiratory system in a patient without the disease or until the diagnosis is proven

R27 Fear of respiratory disease, other *Z71.1*

excl: if patient has the disease code the disease
criteria: excessive fear of other respiratory disease in a patient without the disease or until the diagnosis is proven

R28 Limited function/disability respiratory *Z73.6, Z93.0, Z99.1*

incl: disability due to hypoxia, hypercapnoea, reduced lung function, respiratory diseases, or due to diseases of nose, larynx, or throat
excl: dyspnoea R02, wheezing R03

Note: The COOP/WONCA Charts are suitable for documenting the patient's functional status (see Chapter 8).

R29 Respiratory symptom/complaint, other

R04.1, R04.8, R04.9, R06.6, R09.0, R09.2, R09.8

incl: chest tightness, fluid on lung, hiccough, lung congestion, other symptoms or complaints of respiratory system
excl: pain R01, breathing problems R02, R03, R04, cough R05, nose or sinus symptoms R06, R07, R08, R09, throat symptoms R21, R23, abnormal sputum R24, R25

Component 7—Diagnosis/diseases

(R70) deleted, included with A70

R71 Whooping cough *A37*

incl: parapertussis, pertussis
excl: croup R77
criteria: respiratory infection with a characteristic staccato paroxysmal cough ending with a high-pitched inspiratory whoop, or respiratory infection with cough of at least three weeks' duration in contact with known pertussis, or demonstration of *Bordetella pertussis or parapertussis*
consider: cough R05, upper respiratory infection R74

R72 Strep throat *J02.0, J03.0*

incl: proven streptococcal pharyngitis or tonsillitis
excl: scarlet fever A78; erysipelas, strep skin infection S76
criteria: acute inflammation of the throat, plus demonstration of beta-haemolytic streptococci
consider: tonsillitis R76

R73 Boil, abscess nose *J34.0*

incl: localized nose infection
excl: acute sinusitis R75

R74 Upper respiratory infection, acute *B00.2, B08.5, J00, J02.8, J02.9, J06*

incl: acute rhinitis, coryza, head cold, nasopharyngitis, pharyngitis, upper respiratory tract infection (URTI)
excl: measles A71, infectious mononucleosis A75, viral pharyngoconjunctivitis F70, sinusitis R75, tonsillitis R76, quinsy R76, laryngitis R77, croup R77, influenza R80, allergic rhinitis R97, chronic pharyngitis R83
criteria: evidence of acute inflammation of nasal or pharyngeal mucosa with absence of criteria for more specifically defined acute respiratory infection classified in this section

R75 Sinusitis acute/chronic *J01, J32*

incl: sinusitis affecting any paranasal sinus
criteria: purulent nasal or post-nasal discharge, or previous medically treated episodes of sinusitis, plus tenderness over one or more sinuses, or deep-seated aching facial pain aggravated by dependency of head, or opacity on transillumination; or imaging evidence of sinusitis; or pus obtained from the sinus
consider: headache N01, face pain N03, upper respiratory tract infection R74

R76 Tonsillitis acute *J03.8, J03.9, J36*

incl: peritonsillar abscess, quinsy
excl: infectious mononucleosis A75, strep throat R72, diphtheria R83, hypertrophy or chronic infection of tonsils R90
criteria: sore throat or fever with reddening of tonsil(s) more than the posterior pharyngeal wall, and either pus on swollen tonsil(s) or enlarged tender regional lymph nodes
consider: acute upper respiratory tract infection R74

R77 Laryngitis/tracheitis, acute *J04, J05.0*

incl: croup
excl: laryngotracheobronchitis R78, epiglottitis R83
criteria: hoarseness or stridor with or without respiratory distress, or deep dry painful cough (barking in children), and normal chest signs
consider: upper respiratory tract infection R74

R78 Acute bronchitis/bronchiolitis *J20 to J22, J40*

incl: acute lower respiratory infection NOS; bronchitis NOS; chest infection NOS; laryngotracheobronchitis; tracheobronchitis
excl: influenza R80, chronic bronchitis R79, allergic bronchitis R96
criteria: in children and adults cough and fever with scattered or generalized abnormal chest signs—wheeze, coarse rales, rhonchi, or moist sounds. In infants (bronchiolitis) dyspnoea and hyperinflation
consider: wheezing R03, cough R05, upper respiratory infection R74

R79 Chronic bronchitis *J41, J42*

excl: emphysema, chronic obstructive pulmonary (lung, airways) disease R95; bronchiectasis R99
criteria: cough with sputum on most days for at least 3 months in each of at least 2 years; and scattered rales or rhonchi on auscultation of the chest during these episodes
consider: cough R05; sputum/phlegm abnormal R25; bronchitis NOS R78

R80 Influenza *J10.1, J10.8, J11.1, J11.8*

incl: influenza-like illness, para-influenza
excl: gastric flu D70, influenza pneumonia R81
criteria: myalgia and cough without abnormal respiratory physical signs other than inflammation of nasal mucous membrane and throat, plus three or more of the following—sudden onset (within 12 hours), rigors, chills, or fever, prostration and weakness, influenza in close contacts, or influenza epidemic; or viral culture or serological evidence of influenza virus infection
consider: fever A03; virus infection NOS A77; upper respiratory tract infection R74

R81 Pneumonia *A48.1, J10.0, J11.0, J12 to J16, J18*

incl: bacterial and viral pneumonia; bronchopneumonia; influenzal pneumonia; Legionnaire's disease; pneumonitis
excl: aspiration pneumonia R99
criteria: evidence of pulmonary consolidation
consider: cough R05; bronchitis NOS R78

R82 Pleurisy/pleural effusion *J90, J94, R09.1*

incl: pleural inflammatory exudate, pleuritis
excl: tuberculosis R70, pneumonia R81, malignant effusion code to origin of malignancy
criteria: clinical evidence of pleural exudate; or pleuritic pain accompanied by pleural friction rub; or investigative evidence of inflammatory pleural exudate
consider: pleuritic pain R01

R83 Respiratory infection, other
A36, B37.1, B44, B58.3, J05.1, J31, J37, J85, J86

incl: chronic nasopharyngitis; chronic pharyngitis; chronic rhinitis NOS; fungal respiratory infections; protozoal infections (without pneumonia); lung abscess; diphtheria; epiglottitis; empyema
excl: cystic fibrosis T99

R84 Malignant neoplasm bronchus, lung *C33, C34*

incl: malignancy of trachea, bronchus, lung, or pleura
excl: unknown site A79; secondary malignancy from known site (code to site)
criteria: characteristic histologic appearance
consider: neoplasm of uncertain nature R92

R85 Malignant neoplasm respiratory, other
C09 to C13, C14.0, C14.2, C30.0, C31, C32, C38.4, C39, C45.0

incl: malignancy of larynx, mediastinum, nose, pharynx, sinuses; mesothelioma
excl: Hodgkin's disease B72, malignancy of trachea, bronchus, pleura, or lung R84
criteria: characteristic histologic appearance
consider: neoplasm of uncertain nature R92

R86 Benign neoplasm respiratory *D14, D19*

incl: benign neoplasms of respiratory tract
excl: neoplasm of uncertain nature R92, nasal polyp R99
criteria: characteristic clinical or histologic appearance
consider: neoplasms of uncertain nature R92

R87 Foreign body nose/larynx/bronchus *T17*

incl: foreign body in bronchus, larynx, lung, or nose
excl: drowning A88, foreign body lodged in oesophagus D79, foreign body in ear H76, aspiration pneumonia R99
criteria: visualization of foreign body directly, endoscopically, or using imaging
consider: other complaint of respiratory system R29

R88 Injury respiratory, other *S00.3, S03.1, S27, T27, T70.1*

incl: all injuries and trauma to nose (except fractured nose) and respiratory system; surgical emphysema
excl: drowning A88, fractured nose L76, foreign body R87

R89 Congenital anomaly, respiratory *Q30 to Q34*

incl: congenital abnormality of nose, pharynx, trachea, larynx, bronchi, lungs or pleura
excl: cleft lip or palate D81, cystic fibrosis T99

R90 Hypertrophy tonsils/adenoids *J35*

incl: chronic tonsillitis
excl: acute tonsillitis R76, allergic rhinitis R97

(R91) moved to R79

R92 Neoplasm respiratory, uncertain nature *D02, D38*

incl: uncertain whether benign or malignant
excl: secondary neoplasm unknown site A79; malignant neoplasms R84, R85; benign neoplasm R86

(R93) deleted, included with R82

R95 Chronic obstructive pulmonary disease *J43, J44*

incl: emphysema; chronic obstructive airways (COAD), lung (COLD), pulmonary (COPD) disease; chronic airways limitation (CAL)
excl: chronic bronchitis R79, asthma R96, bronchiectasis R99, cystic fibrosis T99
criteria: objective evidence of airway obstruction, not relieved or only partially relieved by bronchodilators
consider: other breathing problems R04

R96 Asthma *J45, J46*

incl: wheezy bronchitis, reactive airways disease
excl: bronchiolitis R78, chronic bronchitis R79, emphysema R95
criteria: recurrent episodes of reversible acute bronchial obstruction with wheeze and/or dry cough; or diagnostic tests meeting currently accepted criteria for asthma
consider: wheezing R03, cough R05

R97 Allergic rhinitis *J30*

incl: hay fever, nasal allergy, vasomotor rhinitis, hay fever
excl: upper respiratory tract infection R74, rhinitis NOS R83

R98 Hyperventilation syndrome *R06.4*

criteria: symptoms related to hyperventilation and relieved by rebreathing expired air
consider: breathing problems R04

R99 Respiratory disease, other

J33, J34.1 to J34.3, J34.8, J38, J39, J47, J60 to J70, J80 to J82, J84, J92, J93, J96, J98, Z90.2, Z94.2, Z94.3

incl: aspiration pneumonia; bronchiectasis; deviated nasal septum; lung complications of other diseases; mediastinal disease; nasal polyp; other diseases of larynx; pneumoconiosis; pneumothorax; pneumonitis due to allergy, chemicals, dust, fumes, mould; pulmonary collapse; respiratory failure; and other diseases of the respiratory system NOS

S Skin

Component 1—Symptoms and complaints

S01 Pain, tenderness of skin *R20.8, R52*

incl: burning sensation, painful lesion or rash, soreness
excl: tingling, sensation disturbances N05, N06

S02 Pruritus *L29.8, L29.9*

incl: skin irritation
excl: anogenital D05, dermatitis artefacta S99, vulval X16, nipple X20

S03 Warts *B07*

incl: verrucae
excl: molluscum contagiosum S95, genital warts X91, Y76

S04 Lump/swelling, localized *R22*

incl: papule
excl: insect bite S12; breast lump X19

S05 Lumps/swellings, multiple *R21*

incl: papules, lumps or swellings in multiple sites
excl: swollen ankles/oedema K07

S06 Rash localized *L53.9, R21*

incl: blotches, erythema, redness
excl: swellings S05

S07 Rash generalized *L53.9, R21*

incl: blotches, erythema, redness occurring in multiple sites
excl: rash with fever A76, swellings S06

S08 Skin colour change *L81.0, L81.1, L81.2, L81.3, R23.0, R23.1, R23.2*

incl: 'circles under eyes', cyanosis, flushing, freckles, pallor, pigmentation
excl: bruising S16, vitiligo S99

S09 Infected finger/toe *L03.0*

incl: paronychia
excl: post-traumatic S11; tinea S74; monilia, candida S75

S10 Boil/carbuncle *L02*

incl: abscess; furuncle; folliculitis
excl: lymphadenitis B70; perianal D95; external auditory meatus H70; abscess of nose R73; infected finger or toe S09; wound infection S11; erysipelas S76; pilonidal abscess S85; hydradenitis S92; female external genitalia X99; male external genitalia Y99

S11 Skin infection, post-traumatic *T79.3*

incl: infected post-traumatic wound or bite
excl: surgical wound infection A87; erysipelas S76; pyoderma S76; impetigo S84

S12 Insect bite/sting *T14.0*

excl: toxic effects A86; infected bite S11; scabies S72; pediculosis S73

S13 Animal/human bite *T14.1*

excl: toxic effects A86; infected bite S11

S14 Burns/scalds *T20 to T25, T30, T31, T32*

incl: all degrees; chemical burns
excl: sunburn S80

S15 Foreign body in skin *T14.0, T14.1*

incl: foreign body under nail

S16 Bruise/contusion

S00.0, S00.7 to S00.9, S10, S20, S30.0, S30.1, S40, S50, S60, S70, S80, S90, T14.0

incl: haematoma, ecchymosis
excl: with broken skin S17
criteria: intact skin surface

S17 Abrasion/scratch/blister

S00.0, S00.7 to S00.9, S10, S20, S30.7 to S30.9, S40, S50, S60, S70, S80, S90, T14.0

incl: bruise if skin broken; graze

S18 Laceration/cut

S00.0, S00.7 to S00.9, S01.0, S01.2, S01.4, S01.7 to S01.9, S11, S21, S31.0, S31.1, S31.8, S41, S51, S61, S71, S81, S91, T14.1

incl: skin, subcutaneous tissues
excl: bite S13

S19 Skin injury, other

S00.0, S00.7 to S00.9, S10, S20, S30.7 to S30.9, S40, S50, S60, S70, S80, S90, T14.0

incl: avulsion nail, needle stick, puncture
excl: bite S13

S20 Corns/callosities ***L84***

excl: hyperkeratosis S80

S21 Skin texture symptom/complaint ***R23.4***

incl: dry skin, peeling, scaling, wrinkles
excl: sweating problems A09, scalp complaints S24, sweat gland problems S92, ichthyosis S83, vulval complaints X10

S22 Nail symptom/complaint ***L60.1, L60.4, L60.5, L60.9, R68.3***

incl: clubbing
excl: paronychia S09, ingrowing nail S94

S23 Hair loss/baldness ***L63 to L66***

incl: alopecia

S24 Hair/scalp symptom/complaint ***L67, L68***

incl: dry scalp, hirsutism
excl: trichotillomania P29, folliculitis S10, hair loss/baldness S23, dandruff S86

S26 Fear of cancer of skin *Z71.1*

excl: if patient has the disease code the disease
criteria: concern or fear of cancer of skin in a patient without the disease or until cancer is proven

S27 Fear of skin disease, other *Z71.1*

excl: if patient has the disease code the disease
criteria: concern or fear of having other skin disease in a patient without the disease or until the diagnosis is proven

S28 Limited function/disability skin *Z73.6*

incl: disability due to skin condition
Note: The COOP/WONCA Charts are suitable for documenting the patient's functional status (see Chapter 8).

S29 Skin symptom/complaint *R23.3, R23.8*

incl: bleeding skin, cellulite, petechiae, problems with umbilicus, skin lesion, sore(s)
excl: scar S99

Component 7—Diagnosis/diseases

S70 Herpes zoster *B02*

incl: ophthalmic herpes zoster, post-herpetic neuralgia, shingles
criteria: grouped vescicular eruptions, unilateral distribution, over area of a single dermatome
consider: skin pain S01, rash S06

S71 Herpes simplex *B00.0, B00.1, B00.8, B00.9*

incl: cold sore, fever blister
excl: eye F85; genital infections X90, Y72
criteria: vesicles with erythematous base in localized area(s); plus past history of similar lesions, or virological or serological evidence
consider: rash S06

S72 Scabies and other acariases *B86, B88.0, B88.2*

criteria: intense pruritic skin lesions plus arrays of burrows on sides of palms, fingers, penis, or skin folds, or demonstration of parasites or ova in lesions
consider: pruritis S02

S73 Pediculosis/skin infestations, other *B85, B87, B88.1, B88.3, B88.8, B88.9*

incl: fleas, lice, mites, ticks
excl: insect bites S12, infected insect bites S11
criteria: demonstration of nits on hair shafts or insects on skin or clothes
consider: pruritis S02, rash S06

S74 Dermatophytosis *B35, B36*

incl: fungal skin infection, onychomycosis, pityriasis versicolor, ringworm, tinea
excl: monilia, candida S75
criteria: pruritic scaly lesions with central clearing and small vesicles at border; or demonstration of fungus

S75 Moniliasis/candidiasis skin *B37.2*

incl: thrush involving nails, perianal region, skin; monilial intertrigo
excl: oral thrush D83, genital infections X72, Y75

S76 Skin infection, other

A32.0, A46, A66, A67, L03.1 to L03.3, L03.8, L03.9, L08, L98.0

incl: cellulitis, erysipelas, pyoderma, acute lymphangitis
excl: boil, carbuncle, other localized skin infection S11; impetigo S84; molluscum contagiosum S95; acne S96

S77 Malignant neoplasm of skin *C43, C44, C46.0, C46.1*

incl: basal cell carcinoma, malignant melanoma, rodent ulcer, squamous cell carcinoma
excl: premalignant lesions S79
criteria: characteristic histological appearance
consider: other malignant neoplasm (when primary site is uncertain) A79; neoplasm benign/uncertain (when histology is not available) S79

S78 Lipoma *D17*

S79 Neoplasm skin, benign/uncertain *D03, D04, D23, D48.5*

incl: Bowen's disease, dermoid cyst, premalignant lesions
excl: residual haemorrhoidal skin tags K96; hyperkeratoses S99; solar keratoses S80; haemangioma S81; mole S82; pigmented nevus S82; keloid S99; seborrhoeic or senile warts S99

S80 Solar keratosis/sunburn *L55 to L59*

incl: photosensitivity, radiation skin damage, senile keratosis, polymorphous light eruption

(*cont.*)

excl: skin damage due to man made radiation A87 or A88, hyperkeratosis NOS S99

S81 Haemangioma/lymphangioma *D18*

incl: angiomatous birthmark, portwine stain
criteria: vascular or lymphatic tumour, elevated above skin and emptying on pressure
consider: localized swelling S04

S82 Naevus/mole *D22*

S83 Congenital anomaly skin, other *Q80 to Q82, Q84*

incl: birthmark; ichthyosis
excl: haemangioma, lymphangioma S81

S84 Impetigo *L00, L01*

incl: impetigo secondary to other dermatoses
criteria: spreading skin lesion consisting of macules, vesicles, pustules or crust with underlying raw area
consider: other localized skin infection S11

S85 Pilonidal cyst/fistula *L05*

incl: pilonidal abscess
excl: dermoid cyst S79

S86 Dermatitis, seborrhoeic *L21*

incl: cradle cap, dandruff
excl: seborrhoeic warts S99
criteria: greasy, scaly lesions with underlying erythema on one or more areas of scalp, face, sternum, interscapular areas, around umbilicus and in body folds, not attributable to other skin disease
consider: localized rash S06, generalized rash S07

S87 Dermatitis, atopic eczema *L20*

incl: infantile eczema, flexural dermatitis
excl: affecting external auditory meatus alone H70, allergic dermatitis S88, diaper rash S89
criteria: pruritic exudative lesions with or without lichenification over face and neck, wrists and hands, chest, back of knees and front of elbow
consider: rash S06, S07; pruritis S02

S88 Dermatitis, contact/allergic

L23 to L25, L27.2, L27.8, L27.9, L30.0, L30.3, L30.4, L30.8, L30.9

incl: allergic dermatitis, chemical dermatitis, dermatitis NOS, eczema NOS, intertrigo, plant sting, skin allergy
excl: allergy unspecified, allergic reaction unspecified A92; contact and other dermatitis of the eyelid F71; external auditory meatus H70; atopic eczema S87; diaper rash S89; dermatitis artefacta S99; urticaria S98; neurodermatitis S99
criteria: pruritic erythematous lesions related to exposure to chemical substance
consider: pruritis S02; rash S06, S07

S89 Diaper rash — ***L22***

incl: napkin rash
criteria: dermatitis, primarily of the diaper area and sparing creases, in infants

S90 Pityriasis rosea — ***L42***

criteria: oval, scaly eruptions along skin tension lines of trunk, with a history of a solitary lesion preceeding presenting rash
consider: rash S06, S07

S91 Psoriasis — ***L40***

criteria: plaques with silvery scales on knees, elbows or scalp and/or stippled/pitted nails
Note: for psoriatic arthritis double code with L99.

S92 Sweat gland disease — ***L30.1, L73.2, L74, L75***

incl: dyshidrosis; heat rash; hydradenitis; miliaria; prickly heat; sweat rash; pompholyx
excl: hyperhidrosis A09

S93 Sebaceous cyst — ***L72.1***

incl: wen

S94 Ingrowing nail — ***L60.0***

excl: paronychia S09

S95 Molluscum contagiosum — ***B08.1***

S96 Acne — ***L70***

incl: blackheads, comedones, pimples
excl: due to medications A85

S97 Chronic ulcer skin *I83.0, I83.2, L89, L97, L98.4*

incl: bedsore, pressure sore, varicose ulcer, decubitus ulcer
excl: gangrene K92

S98 Urticaria *L50*

incl: hives, weals
excl: drug allergy A85; angioedema, allergic oedema A92

S99 Skin disease, other

L10 to L13, L26, L28, L30.2, L30.5, L41, L43 to L44, L51, L52, L53.0 to L53.3, L53.8, L60.2, L60.3, L60.8, L71, L72.0, L72.2, L72.8, L72.9, L73.0, L73.1, L73.8, L73.9, L80, L81.4 to L81.9, L82, L83, L85, L87, L88, L90 to L95, L98.1 to L98.3, L98.5 to L98.9, Z94.5

incl: dermatitis artefacta; discoid lupus erythematosus; erythema multiforme; erythema nodosum; granuloma; granuloma annulare; hyperkeratosis NOS; keloid; keratoacanthoma; lichen planus; neurodermatitis; onychogryphosis; rosacea; rhinophyma; scar; seborrhoeic or senile warts; striae atrophicae; vitiligo; and other diseases of the skin NOS

T Endocrine, metabolic, and nutritional

Component 1—Symptoms and complaints

T01 Excessive thirst *R63.1*

incl: polydipsia

T02 Excessive appetite *R63.2*

incl: overeating, polyphagia
excl: bulimia P86

T03 Loss of appetite *R63.0*

incl: anorexia
excl: anorexia nervosa P86

T04 Feeding problem of infant/child *R63.3*

incl: problem of what and how to eat/feed
excl: food allergy A98; psychological causes P11; eating disorders P11; food intolerance T99

T05 Feeding problem of adult *R63.3*

incl: problem of what and how to eat/feed
excl: food allergy A98; dysphagia D21; psychological eating disorders, food refusal P29; anorexia/bulimia nervosa P86; loss of appetite T03; food intolerance T99

(T06) deleted, transferred to P86

T07 Weight gain *R63.5*

excl: obesity T82, overweight T83

T08 Weight loss *R63.4, R64*

incl: cachexia
excl: anorexia nervosa P86

T10 Growth delay *E34.3, R62.8, R62.9*

incl: failure to thrive, physiological delay
excl: delayed milestones P22; learning disorder P24; mental retardation P85; delayed puberty T99

T11 Dehydration *E86*

incl: water depletion
excl: salt depletion T99, electrolyte disturbance T99

(T15) deleted, included with T81

T26 Fear of cancer of endocrine system *Z71.1*

excl: if patient has the disease code the disease
criteria: concern or fear of cancer of endocrine system in a patient without the disease or until the diagnosis is proven

T27 Fear of endocrine/metabolic disease, other *Z71.1*

incl: fear of diabetes
excl: if patient has the disease code the disease
criteria: concern or fear of other endocrine, metabolic, or nutritional disease in a patient without the disease or until the diagnosis is proven

T28 Limited function/disability endocrine/metabolic *Z73.6*

incl: limited function/disability related to the endocrine system or metabolic or nutritional factors

Note: The COOP/WONCA Charts are suitable for documenting the patient's functional status (see Chapter 8).

T29 Endocrine/metabolic/nutritional symptom/complaint, other *R63.8*

incl: specific food craving, underweight
excl: hyperglycaemia A91; fluid retention K07

Component 7—Diagnosis/diseases

T70 Endocrine infection *E06.0*

excl: thyroiditis T99

T71 Malignant neoplasm thyroid *C73*

criteria: characteristic histological appearance
consider: goitre T81, unspecified neoplasm T73

T72 Benign neoplasm thyroid *D34*

excl: goitre T81, unspecified neoplasm T73

T73 Neoplasm endocrine other/uncertain *C74, C75, D09.3, D35, D44*

T78 Thyroglossal duct/cyst *Q89.2*

excl: goitre T81

T80 Congenital anomaly endocrine/metabolic *E00, Q89.1, Q89.2*

incl: dwarfism, cretinism
excl: thyroglossal duct (cyst) T78

T81 Goitre *E04*

incl: thyroid nodule, non-toxic goitre
excl: neoplasm of thyroid gland T71–T72, thyroglossal cyst T78, toxic goitre T85, hypothyroidism T86

T82 Obesity *E66*

excl: overweight T83
criteria: a body mass index greater than 30

T83 Overweight *E66*

excl: obesity T82
criteria: a body mass index greater than 25 but less than 30

T85 Hyperthyroidism/thyrotoxicosis ***E05***

incl: Graves disease, toxic goitre
excl: non-toxic goitre T81
criteria: laboratory evidence of excessive thyroid hormone; or thyroid nodule or goitre plus tremor, weight loss, and rapid pulse (over 100/min at rest) or eye signs (exophthalmos, lid lag, or ophthamoplegia)

T86 Hypothyroidism/myxoedema ***E01 to E03***

excl: cretinism T80
criteria: laboratory evidence of diminished thyroid hormone activity and excessive thyroid stimulating hormone; or four or more of: weakness, tiredness; mental changes: apathy, poor memory, slowing; voice changes: coarser, deeper slower speech; undue sensitivity to cold; constipation; coarse puffy facial features; cool dry, sallow skin, decreased sweating; peripheral oedema
consider: other complaint of metabolism T29

T87 Hypoglycaemia ***E15, E16***

incl: hyperinsulinism, insulin coma
criteria: hypoglycaemia demonstrated by biochemical testing, or characteristic symptoms in a diabetic patient relieved by ingestion or injection of sugar

(T88) deleted, included with T99

T89 Diabetes, insulin dependent ***E10***

incl: juvenile onset diabetes, type 1 diabetes
excl: gestational diabetes W85
criteria: patient requiring regular ongoing treatment with insulin after diagnosis confirmed by one of the following:

(a) the classic symptoms of diabetes, such as polyuria, polydypsia, and rapid weight loss, together with unequivocal elevation of plasma glucose
(b) fasting blood glucose levels of 8 mmol/l (140 mg/dl) or more on two or more occasions
(c) random blood glucose levels of 11 mmol/l (200 mg/dl) or more on two or more occasions
(d) an oral Glucose Tolerance Test (75 gm glucose) with one value of plasma glucose at between 1 and 2 hours of 11 mmol/l (200 mg/dl) or more, and plasma glucose at 2 hours of 11 mol/l (200 mg/dl) or more

consider: hyperglycaemia A91; diabetes, non-insulin dependent T90
Note: 1. double code complications such as retinopathy F83, nephropathy U88.
2. in pregnancy double code with W84.

T90 Diabetes, non-insulin dependent *E11 to E14*

incl: diabetes NOS, late onset diabetes, type 2 diabetes
excl: drug induced hyperglycaemia A85, gestational diabetes W85, hyperglycaemia as isolated finding A91
criteria: patient not requiring regular ongoing treatment with insulin after diagnosis confirmed by one of the following:

(a) the classic symptoms of diabetes, such as polyuria, polydypsia, and rapid weight loss, together with unequivocal elevation of plasma glucose
(b) fasting blood glucose levels of 8 mmol/l (140 mg/dl) or more on two or more occasions
(c) random blood glucose levels of 11 mmol/l (200 mg/dl) or more on two or more occasions
(d) an oral Glucose Tolerance Test (75 g glucose) one value of plasma glucose at between 1 and 2 hours of 11mmol/l (200 mg/dl) or more and plasma glucose at 2 hours of 11 mol/l (200 mg/dl) or more

consider: hyperglycaemia A91; diabetes, insulin dependent T89
Note: 1. double code complications such as retinopathy F83, nephropathy U88.
2. in pregnancy double code with W84.

T91 Vitamin/nutritional deficiency

E40 to E46, E50, E51.1, E51.8, E51.9, E52 to E56, E58 to E61, E63, E64

incl: beri-beri; dietary mineral deficiency; iron deficiency without anaemia; malnutrition; marasmus; scurvy
excl: iron deficiency anaemia B80; megaloblastic (pernicious) anaemia B81; malabsorption syndrome, sprue D99

T92 Gout *M10*

excl: drug induced gout A85; raised uric acid A91; pseudo-gout and other crystal arthropathies T99

T93 Lipid disorder *E78*

incl: hyperlipidaemia; abnormalities of lipoprotein levels; raised levels of cholesterol and triglycerides; xanthoma

T99 Endocrine/metabolic/nutritional disease, other

E06.1 to E06.5, E06.9, E07, E20 to E32, E34.0 to E34.2, E34.4 to E34.9, E65, E67, E68, E70 to E77, E79, E80, E84, E85, E87 to E88, M11, M83

incl: acromegaly; amyloidosis; congenital metabolic disorders; crystal arthropathy; Cushing's syndrome; cystic fibrosis; diabetes insipidus; Gilbert's syndrome; hyperaldosteronism; lactose intolerance; osteomalacia; adrenal/

(*cont.*)

ovarian/pituitary/parathyroid/testicular or other endocrine dysfunction; porphyria; precocious or delayed puberty; pseudo-gout; renal glycosuria; thyroiditis

excl: food allergy A98; food intolerance D99; osteoporosis L95

U Urinary system

Component 1—Symptoms and complaints

U01 Dysuria/painful urination ***R30***

incl: burning urination
excl: frequent/urgent urination U02, urethral syndrome U72

U02 Urinary frequency/urgency ***R35***

incl: nocturia, polyuria

U04 Incontinence urine ***N39.3, N39.4, R32***

incl: enuresis of organic origin, involuntary urination, stress incontinence
excl: of psychogenic origin P12

U05 Urination problems, other ***R34, R39.1***

incl: anuria, oliguria, dribbling urine
excl: urinary retention U08

U06 Haematuria ***N02, R31***

incl: blood in urine, macroscopic, microscopic, or positive chemical test

U07 Urine complaints, other ***R39.8***

incl: bad odour of urine, dark urine
excl: abnormal urine test U98

U08 Urinary retention ***R33***

U13 Bladder symptom/complaint, other ***R39.0, R39.8***

incl: irritable bladder

U14 Kidney symptom/complaint ***N23***

incl: kidney pain, kidney trouble, renal colic
excl: loin/flank pain L05

U26 Fear of cancer of urinary system *Z71.1*

excl: if patient has the disease code the disease
criteria: concern or fear of cancer of urinary system in a patient without the disease or until the disease is proven

U27 Fear of urinary disease, other *Z71.1*

excl: fear of cancer of urinary system U26; if patient has the disease code the disease
criteria: concern or fear of other urinary disease in a patient without the disease or until the diagnosis is proven

U28 Limited function/disability urinary *Z73.6, Z93.5, Z93.6, Z99.2*

incl: renal transplant, slow stream
excl: urinary incontinence U04
Note: The COOP/WONCA Charts are suitable for documenting the patient's functional status (see Chapter 8).

U29 Urinary symptom/complaint, other *R39.8*

excl: irritable bladder, bladder pain U13; kidney problems U14

Component 7—Diagnosis/diseases

U70 Pyelonephritis/pyelitis *N10 to N12, N15.1, N15.9*

incl: infection of kidney; renal and perinephric abscess
criteria: two of flank pain, or renal tenderness, or investigation evidence of chronic renal damage; plus clinical or laboratory evidence of urinary tract infection
consider: cystitis/other urinary infection U71

U71 Cystitis/urinary infection, other *N30, N39.0*

incl: asymptomatic bacteriuria, acute and chronic cystitis (non-venereal), lower urinary tract infection, urinary tract infection NOS
excl: pyelonephritis U70, urethritis U72; vaginitis X84; balanitis Y75;
Note: in pregnancy also code W84.

U72 Urethritis *B37.4, N34*

incl: chlamydial urethritis, non-specific urethritis, urethral syndrome, meatitis
excl: gonococcal urethritis female X71; urethritis trichomonal, female X73; gonococcal urethritis male Y71
criteria: urethral discharge with frequency, burning, pain or urgency on urination without bacteruria by microscopy or culture; or inflammation of external urinary meatus

(cont.)

consider: painful urination U01; frequent/urgent urination U02; irritable bladder U13; urethral discharge X29, Y03

U75 Malignant neoplasm kidney ***C64, C65***

criteria: characteristic histologic appearance
consider: neoplasm urinary tract NOS U79 (when histology not available)

U76 Malignant neoplasm bladder ***C67***

criteria: characteristic histologic appearance
consider: neoplasm urinary tract NOS U79 (when histology not available)

U77 Malignant neoplasm, urinary, other ***C66, C68***

incl: malignant neoplasm ureter, malignant neoplasm urethra
excl: malignant neoplasm prostate Y77
criteria: characteristic histologic appearance
consider: neoplasm urinary tract U79 NOS (when histology not available)

U78 Benign neoplasm urinary tract ***D30***

incl: bladder papilloma; polyp of urinary tract
excl: prostatic hypertrophy Y85
criteria: characteristic histologic appearance
consider: neoplasm urinary tract U79 NOS (when histology not available)

U79 Neoplasm urinary tract NOS ***D09.0, D09.1, D41***

incl: bladder/kidney/ureter/urethra neoplasm NOS
excl: histologically proven urinary tract neoplasms U75, U76, U77, U78

U80 Injuries urinary tract ***S37.0 to S37.3, T19.0, T19.1, T28.3, T28.8***

incl: contusion kidney, foreign body in urinary tract

U85 Congenital anomaly urinary tract ***Q60 to Q64***

incl: duplex kidney/ureter, polycystic kidney (congenital)

U88 Glomerulonephritis/nephrosis
N00, N01, N03, N04, N05, N07, N14, N15.0, N15.8

incl: acute glomerulonephritis, analgesic nephropathy, chronic glomerulonephritis, nephritis, nephropathy, nephrotic syndrome
excl: renal failure U99
criteria: three or more of: hematuria, proteinuria, renal salt and water retention, decreased renal function, persistent urinary sediment abnormalities; or renal biopsy evidence

(*cont.*)

consider: abnormal urine test U98, kidney complaint U14

U90 Orthostatic albuminuria/proteinuria *N39.2*

incl: postural proteinuria
criteria: albuminuria following ambulation, no albuminuria following overnight recumbency and no evidence of renal disease
consider: proteinuria NOS U98

U95 Urinary calculus *N20, N21*

incl: calculus or stone in bladder, kidney, ureter; urolithiasis
criteria: colicky pain and either haematuria or history of urinary stone in the past; or passage of calculus; or imaging evidence of calculus
consider: blood in urine U06, renal colic U14, other urinary symptoms U29, abnormal urine test finding U98

U98 Abnormal urine test NOS *N39.1, R80 to R82*

incl: glycosuria, proteinuria, pus in urine, pyuria
excl: haematuria/blood in urine U06; orthostatic albuminuria/proteinuria U90

U99 Urinary disease, other

N06, N13, N17 to N19, N25 to N28, N31, N32, N35 to N36, N39.8, N39.9, R39.2, T19.8, T19.9, Z90.5, Z90.6, Z94.0

incl: bladder diverticulum; hydronephrosis; hypertrophic kidney; obstruction bladder neck; renal failure; urethral caruncle; urethral stricture; ureteric reflux; uraemia; and other diseases of the urinary system NOS

W Pregnancy, childbearing, family planning

Component 1—Symptoms and complaints

W01 Question of pregnancy *Z32.0*

incl: delayed menstruation or symptoms suggestive of pregnancy
excl: fear of pregnancy W02, pregnancy confirmed W78,W79

W02 Fear of pregnancy *Z71.1*

incl: concern about possibility of unwanted pregnancy
excl: concern or fear if unwanted pregnancy confirmed W79

W03 Antepartum bleeding *O20, O46*

incl: vaginal haemorrhage in pregnancy
excl: intrapartum hemorrhage W17

W05 Pregnancy nausea/vomiting *O21*

incl: hyperemesis, morning sickness in confirmed pregnancy

W10 Contraception, post-coital *Z30.3*

incl: 'morning after pill'

W11 Contraception, oral *Z30.4*

incl: family planning using hormonal therapy

W12 Contraception, intrauterine *T83, Z30.1, Z30.5*

incl: family planning using intrauterine device IUD

W13 Sterilization *Z30.2*

incl: family planning involving female sterilization

W14 Contraception, other *Z30.0, Z30.8, Z30.9*

incl: contraception NOS, family planning NOS, genetic counselling
excl: oral contraception W11, IUD W12, sterilization W13

W15 Infertility/subfertility *N97, Z31*

incl: sterility, primary and secondary
criteria: failure to conceive after 2 years of trying
consider: symptom/complaint about pregnancy W29

W17 Post-partum bleeding *O72*

incl: post-partum haemorrhage
criteria: heavy bleeding at or within 6 weeks of parturition
consider: post-partum complaints W18

W18 Post-partum symptom/complaint *O90.9*

incl: complaints relating to and within 6 weeks of parturition
excl: puerperal depression P76, bleeding W17, complaints about lactation W19, complications W96

W19 Breast/lactation symptom/complaint *O92.5 to O92.7*

incl: galactorrhoea, suppression of lactation, weaning
excl: puerperal mastitis W94, cracked nipples W95

(W20) deleted, included with W19

W21 Concern about body image related to pregnancy *R46.8*

W27 Fear of complications of pregnancy *Z71.1*

incl: fear of congenital anomalies in baby
excl: if patient has the complication code the complication
criteria: concern or fear of complications in a patient without them or until they are proven

W28 Limited function/disability pregnancy *Z73.6*

incl: disability due to or related to pregnancy, pelvic instability
Note: The COOP/WONCA Charts are suitable for documenting the patient's functional status (see Chapter 8).

W29 Pregnancy symptom/complaint, other *R68.8*

incl: family planning symptom, complaint

Component 7—Diagnoses/diseases

W70 Puerperal infections/sepsis *O85, O86.1, O86.3*

excl: obstetric tetanus N72
criteria: infection of birth canal or reproductive organs within 6 weeks of parturition

W71 Infections complicating pregnancy
O23, O41.1, O75.2, O75.3, O86.2, O86.4, O86.8, O98

incl: maternal infections complicating pregnancy or the puerperium
excl: puerperal infection W70

W72 Malignant neoplasm related to pregnancy *C58*

incl: chorioepithelioma, choriocarcinoma

W73 Benign/uncertian neoplasm related to pregnancy *O01*

incl: hydatidiform mole

W75 Injury complicating pregnancy *T14.9*

incl: results of injury which interfere with pregnancy
excl: new injury caused by childbirth W92, W93

W76 Congenital anomaly complicating pregnancy *O99.8*

incl: maternal anomalies which could affect pregnancy, childbirth

W78 Pregnancy ***Z32.1, Z33, Z34***

incl: pregnancy confirmed
excl: unwanted pregnancy W79, ectopic pregnancy W80, high-risk pregnancy W84

W79 Unwanted pregnancy ***Z32.1***

W80 Ectopic pregnancy ***O00***

criteria: confirmation by palpation, ultrasound, laparoscopy, culdoscopy or surgery
consider: antepartum bleeding W03, other symptoms in pregnancy W29

W81 Toxaemia of pregnancy ***O10 to O16***

incl: eclampsia; hypertension, oedema, proteinuria in pregnancy; pre-eclampsia
criteria: diastolic blood pressure >90 mmHg on at least two occasions or rises 15 mmHg or more; or oedema; or proteinuria; or convulsions with hypertension or oedema or proteinuria
consider: other symptom or complaint in pregnancy W29

W82 Abortion, spontaneous ***O02, O03, O05, O06***

incl: miscarriage; abortion—threatened, complete, incomplete, missed, habitual
excl: induced abortion W83, antepartum bleeding W03, premature contractions after the 22nd week of pregnancy W92, W93, foetal death, stillbirth after the 22nd week of pregnancy W93

W83 Abortion, induced ***O04, Z30.3***

incl: termination of pregnancy, all complications

W84 Pregnancy high risk
O24.0 to O24.3, O24.9, O25, O30 to O36, O40, O43, O44, O99, Z35

incl: aged primipara; anaemia of pregnancy; diabetes or other pre-existing chronic disease affecting pregnancy (see note); disproportion; hydramnios; malpresentation; multiple pregnancy; placenta praevia; previous caesarian section; small foetus for age
excl: infections complicating pregnancy W71, ectopic pregnancy W80, toxaemia of pregnancy W81, gestational diabetes W85
Note: double code diabetes (T89 or T90) or other pre-existing chronic disease.

W85 Gestational diabetes ***O24.4***

incl: diabetes manifested during pregnancy
excl: pre-existing diabetes T89 or T90

(*cont.*)

criteria: fasting plasma glucose over 5.5 mmol/l and/or plasma glucose level greater than 8.0 mmol/l 2 hours after a 75 g oral glucose tolerance test
consider: hyperglycaemia A91

W90 Uncomplicated delivery, livebirth *O80, Z37.0, Z37.9, Z38*

incl: labour, normal delivery NOS

W91 Uncomplicated delivery, stillbirth *Z37.1, Z37.9*

incl: labour

W92 Complicated delivery, livebirth

O42, O45, O60 to O71, O73, O75.0, O75.1, O75.4 to O75.9, O81 to O84, Z37.2, Z37.5, Z37.9, Z38

incl: assisted extraction; breech delivery; caesarian section; dystocia; induction of labour; injuries caused by childbirth; placenta praevia in delivery; version
excl: haemorrhage W17, eclampsia W81

W93 Complicated delivery, stillbirth

O42, O45, O60 to O71, O73, O75.0, O75.1, O75.4 to O75.9, O81 to O84, Z37.1, Z37.3, Z37.4, Z37.6, Z37.7, Z37.9

incl: assisted extraction; breech delivery; caesarian section; dystocia; induction of labour; injuries caused by childbirth; placenta praevia in delivery; version
excl: haemorrhage W17, eclampsia W81

W94 Puerperal mastitis *O91*

incl: breast abscess
criteria: pain, inflammation of breast within 6 weeks of parturition or while lactating
consider: disorders of lactation W19

W95 Breast disorder in pregnancy/puerperium, other

O92.0 to O92.4

incl: breast disorder in puerperium, cracked nipple
excl: disorders of lactation W19; mastitis W94; disorder not related to pregnancy or lactation X21

W96 Complications of puerperium, other *O87, O90.4, O90.5, O90.8, O90.9*

excl: puerperal depression P76, puerperal infection W70, toxaemia W81, breast disorders W95

W99 Disorder of pregnancy/delivery, other

O07, O08, O22, O26, O28, O41.0, O41.8, O41.9, O47, O48, O88, O95 to O97

excl: pseudocyesis P75

X Female genital system (including breast)

Component 1—Symptoms and complaints

X01 Genital pain female *R52*

incl: pelvic pain, vulval pain
excl: menstrual pain X02, dyspareunia X04, breast pain X18

X02 Menstrual pain *N94.4 to N94.6*

incl: dysmenorrhoea

X03 Intermenstrual pain *N94.0*

incl: mittelschmerz, ovulation pain

X04 Painful intercourse *N94.1, N94.2*

incl: dyspareunia, vaginismus NOS
excl: psychogenic sexual problems P07, P08

X05 Menstruation absent/scanty *N91*

incl: amenorrhoea, delayed or late menses, oligomenorrhoea
excl: question of pregnancy W01, fear of pregnancy W02

X06 Menstruation excessive *N92.0, N92.2, N92.4*

incl: menorrhagia, pubertal bleeding

X07 Menstruation irregular/frequent *N92.1, N92.5, N92.6*

incl: polymenorrhoea
excl: menorrhagia X06, pubertal bleeding X08

X08 Intermenstrual bleeding *N92.3, N93.8, N93.9*

incl: breakthrough bleeding, dysfunctional uterine bleeding, metrorrhagia, ovulation bleeding, spotting
excl: post-menopausal bleeding X12, post-coital bleeding X13

X09 Premenstrual symptom/complaint *N94.9*

incl: any symptom related to premenstrual period
excl: premenstrual tension syndrome X89

X10 Postponement of menstruation *Z30.9*

criteria: postponement of the expected regular menstruation by hormonal treatment

Sexual Transmitted Diseases can cause Infertility to female

X11 Menopausal symptom/complaint *N95.1 to N95.3, N95.8, N95.9*

incl: symptoms related to menopause, menopause syndrome, atrophic vaginitis, senile vaginitis
excl: post-menopausal bleeding X12

X12 Post-menopausal bleeding *N95.0*

criteria: vaginal bleeding following either at least 6 months amenorrhoea or demonstration of menopause by appropriate laboratory test
consider: irregular menstruation X07

X13 Post-coital bleeding *N93.0*

incl: contact bleeding

X14 Vaginal discharge *N89.8*

incl: fluor vaginalis, leukorrhoea
excl: bleeding X06–X08, atrophic vaginitis X11, gonorrhoea X71, urogenital, candidiasis X72, urogenital trichomoniasis X73, chlamydia X92

X15 Vaginal symptom/complaint, other *N89.9*

incl: dryness
excl: genital pain X01, vaginismus organic X04, atrophic vaginitis X11

X16 Vulval symptom/complaint *L29.2, N90.9*

incl: pruritis, dryness
excl: vulval pain X01, abscess vulva X99

X17 Pelvis symptom/complaint *N94.9*

excl: genital pain X01

X18 Breast pain *N64.4*

incl: mastodynia
excl: in pregnancy or lactation W19

X19 Breast lump/mass *N63*

incl: lumpy breasts

X20 Nipple symptom/complaint *N64.0, N64.5*

incl: nipple discharge, fissure, pain, pruritis, retraction
excl: in pregnancy or lactation W19

X21 Breast symptom/complaint, other *N61, N62, N64.3, N64.5, N64.9*

incl: mastitis (non-lactating), mastopathy, galactorrhoea
excl: mastitis (lactating) W94

X22 Concern about breast appearance *R46.8*

X23 Fear of sexually transmitted disease *Z71.1*

excl: if patient has the disease code the disease
criteria: concern or fear of sexually transmitted disease in a patient without the disease or until the diagnosis is proven

X24 Fear of sexual dysfunction female *Z70, Z71.1*

excl: if patient has sexual dysfunction P07 or P08
criteria: concern or fear of sexual dysfunction in a patient without sexual dysfunction or until the diagnosis is proven

X25 Fear of genital cancer female *Z71.1*

excl: if patient has the disease code the disease
criteria: concern or fear of genital cancer in a patient without the disease or until the diagnosis is proven

X26 Fear of breast cancer *Z71.1*

excl: if patient has the disease code the disease
criteria: concern or fear of breast cancer in a patient without the disease or until the diagnosis is proven

X27 Fear genital/breast disease, other female *Z71.1*

excl: if patient has the disease code the disease
criteria: concern or fear of other genital/breast disease in a patient without the disease or until the diagnosis is proven

X28 Limited function/disability genital, female *Z73.6, Z90.7*

excl: sexual dysfunction P07, P08; painful intercourse X04, vaginismus X04
Note: The COOP/WONCA Charts are suitable for documenting the patient's functional status (see Chapter 8).

X29 Genital symptom/complaint, other female *N94.9, R36*

incl: urethral discharge in female

Component 7—Diagnosis/diseases

X70 Syphilis, female *A50 to A53, A65*

incl: syphilis any site
criteria: demonstration of *Treponema pallidum* on microscopy, or positive serological test for syphilis

X71 Gonorrhoea, female *A54*

incl: gonorrhoea any site
criteria: purulent urethral or vaginal discharge in a contact with a proven case; or gram negative intra-cellular diplococci demonstrated in discharge; or culture of *Neisseria gonorrhoea*
consider: urethritis U72; urethral discharge X29

X72 Genital candidiasis *B37.3, B37.4*

incl: monilial infection of vagina or cervix; thrush
criteria: inflamed urogenital mucosa or skin with characteristic white adherent exudate or demonstration of candida
consider: vaginal discharge X14; vaginitis X84

X73 Genital trichomoniasis *A59.0*

criteria: characteristic foul smelling discharge or demonstration of trichomonads on microscopy
consider: vaginal discharge X14; vaginitis X84

X74 Pelvic inflammatory disease *N70, N71, N73*

incl: endometritis; oophoritis; salpingitis
excl: venereal diseases X70–X73; chlamydia infection X92
criteria: lower abdominal pain with marked tenderness of uterus or adnexa by palpation, plus other evidence of inflammation
consider: pelvic congestion syndrome X99

X75 Malignant neoplasm cervix *C53, D06*

excl: carcinoma-in-situ X81; cervical intraepithelial neoplasia (CIN) grade 3 X81; abnormal Pap smear X86; CIN grades 1 and 2 X86
criteria: characteristic histological appearance
consider: other malignant neoplasms (when primary site is uncertain) X77; neoplasms when histology is not available X81

X76 Malignant neoplasm breast *C50*

incl: intraductal carcinoma

(cont.)

excl: exclude carcinoma-in-situ X81
criteria: characteristic histological appearance
consider: breast lump X19

X77 Malignant neoplasm genital, other *C51, C52, C54 to C57*

incl: malignancy of adnexae, ovaries, uterus, vagina, vulva
excl: carcinoma in-situ X81
criteria: characteristic histological appearance
consider: genital neoplasm X81 when histology not available

X78 Fibromyoma uterus *D25*

incl: fibroid uterus, fibromyoma of cervix; myoma
excl: cervical polyp (adenomatous) X80, mucous cervical polyp X85
criteria: enlargement of the uterus not due to pregnancy or malignancy, with single or multiple firm tumours of the uterus

X79 Benign neoplasm breast *D24*

incl: fibroadenoma
excl: cystic disease of breast X88
criteria: characteristic histological appearance
consider: lump in breast X19

X80 Benign neoplasm, genital female *D26 to D28*

excl: polyp of cervix X85; physiological cyst of ovary X99

X81 Genital neoplasm, other/uncertain
D05, D06, D07.0 to D07.3, D39, D48.6

incl: carcinoma-in-situ, biopsy proven cervical intraepithelial neoplasia (CIN) grade 3
excl: endometrial polyp X99

X82 Injury genital
S30.2, S31.4, S31.5, S37.4 to S37.6, S38.0, S38.2, T19.2, T19.3, T28.3, T28.8

incl: foreign body in vagina, female circumcision
excl: due to childbirth W92, W93

X83 Congenital anomaly genital female *Q50 to Q52, Q56, Q83*

incl: hermaphroditism, imperforate hymen
excl: other genetic syndromes A90

X84 Vaginitis/vulvitis NOS *N76*

incl: vaginosis, gardnerella

(*cont.*)

excl: genital candidiasis X72; genital trichomoniasis X73; atrophic vaginitis X11

X85 Cervical disease NOS *N72, N84.1, N86 to N88*

incl: biopsy proven CIN 1, CIN 2 and cervical dysplasia; cervical erosion; cervical leukoplakia; cervicitis; mucous cervical polyp; old laceration
excl: abnormalities of cervix in pregnancy, childbirth, and puerperium W77; adenomatous polyp X78; abnormal Pap smear NOS X86

X86 Abnormal pap smear *R87*

incl: cervical intraepithelial neoplasia (CIN) grades 1, 2 and 3, and dysplasia, not proven by biopsy
excl: biopsy proven cervical intraepithelial neoplasia (CIN) grades 1 and 2 and dysplasia X85, grade 3 X81, biopsy proven carcinoma-in-situ X81

X87 Uterovaginal prolapse *N81*

incl: cystocoele; rectocoele; procidentia
excl: stress incontinence U04

X88 Fibrocystic disease breast *N60*

incl: chronic cystic disease of breast, cystic fibroadenosis, dysplasia, solitary cyst of breast

X89 Premenstrual tension syndrome *N94.3*

criteria: cyclic occurrence in the menstrual cycle of two or more of oedema, breast tenderness or swelling, headache, irritability, mood changes
consider: premenstrual symptom X09

X90 Genital herpes *A60*

incl: ano-genital herpes simplex

X91 Condylomata acuminata *A63.0*

incl: venereal warts, human papilloma virus infection

X92 Chlamydia infection, genital *A56*

X99 Genital disease, other
A55, A57, A58, A63.8, N61, N64.1, N64.2, N64.8, N75, N80, N82, N83, N84.0, N84.2, N84.3, N84.8, N84.9, N85, N89.0 to N89.7, N90.0 to N90.8, N94.8, N94.9, N96, N98, Z90.1, Z90.7

incl: Bartholin cyst or abscess; endometrial polyp; endometriosis; genital tract fistula; pelvic congestion syndrome; physiological ovarian cyst (*cont.*)

excl: sexually transmitted disease NOS A78

Y Male genital system

Component 1—Symptoms and complaints

Y01 Pain in penis *N48.9, R52*

excl: priapism, painful erection Y08

Y02 Pain in testis/scrotum *R10.2, N50.9, R52*

incl: perineum, pelvis

Y03 Urethral discharge *R36*

Y04 Penis symptom/complaint *N48.9*

incl: foreskin symptom/complaint
excl: priapism Y08

Y05 Scrotum/testis symptom/complaint *L29.1, N50.9*

incl: lump in testis

Y06 Prostate symptom/complaint *N40, N42.9,*

incl: prostatism
excl: urinary retention U08

Y07 Impotence NOS *N48.4*

incl: impotence of organic origin
excl: reduced sexual desire P07, reduced sexual fulfilment P08, psychogenic impotence P08

Y08 Sexual function symptom/complaint *N48.3, N48.9*

incl: priapism
excl: reduced sexual desire P07, reduced sexual fulfilment P08, psychogenic impotence P08

Y10 Infertility/subfertility *N46*

criteria: failure of conception after 2 years of trying

Y13 Sterilization *Z30.2*

incl: family planning involving male sterilization, vasectomy

Y14 Family planning, other *Z30.0, Z30.8, Z30.9*

incl: contraception NOS, family planning NOS, genetic counselling

Y16 Male breast symptom/complaint *N62, N63*

incl: lump male breast, gynaecomastia
excl: disease of male breast Y99

Y24 Fear of sexual dysfunction *Z71.1*

excl: in a patient with the disease code the disease
criteria: concern about or fear of sexual dysfunction in a patient without the condition

Y25 Fear of sexually transmitted disease *Z71.1*

excl: fear of HIV or AIDS B25; in a patient with the disease code the disease
criteria: concern about or fear of venereal disease in a patient without the disease or until the diagnosis is proven

Y26 Fear of cancer male genital *Z71.1*

excl: in a patient with the disease code the disease
criteria: concern about or fear of cancer in a patient without the disease or until the diagnosis is proven

Y27 Fear of genital disease, other *Z71.1*

excl: in a patient with the disease code the disease
criteria: concern about or fear of other genital disease in a patient without the disease or until the diagnosis is proven

Y28 Limited function/disability genital, male *Z73.6, Z90.7*

excl: sexual dysfunction P07, P08
Note: The COOP/WONCA Charts are suitable for documenting the patient's functional status (see Chapter 8).

Y29 Genital symptom/complaint other, male *N50.9*

Component 7—Diagnosis/diseases

Y70 Syphilis male *A50 to A53, A65*

incl: syphilis any site

(*cont.*)

criteria: demonstration of *Treponema pallidum* on microscopy, or positive serological test for syphilis

Y71 Gonorrhoea male ***A54***

incl: gonorrhoea any site
criteria: urethral or rectal discharge with gram-negative intracellular diplococci demonstrated or *Neisseria gonorrhoea* cultured
consider: urethritis U72; urethral discharge Y03

Y72 Genital herpes, male ***A60***

incl: anogenital herpes; herpes genitalis male
criteria: small vesicles with characteristic appearance and location which evolve to painful ulcers and scabs

Y73 Prostatitis/seminal vesiculitis ***A59.0, N41, N49.0***

criteria: tenderness of prostate or seminal vesicles to palpation, and indications of inflammation in urine test

Y74 Orchitis/epididymitis ***N45***

excl: mumps D71; tuberculosis A70; gonococcal Y71; torsion of testis Y99
criteria: both swelling and tenderness of testes or epididymis, and absence of a specific aetiology (mumps, gonococcal, tuberculosis, trauma, or torsion)
consider: symptoms of testis Y05

Y75 Balanitis ***B37.4, N48.1***

incl: candidiasis
excl: syphilis Y70, gonorrhoea Y71, herpes Y72, scabies S72
criteria: signs of inflammation of the prepuce or glans penis

Y76 Condylomata acuminata, male ***A63.0***

incl: venereal warts; human papilloma virus infection
criteria: characteristic appearance of lesions, or characteristic histological appearance

Y77 Malignant neoplasm prostate ***C61***

criteria: characteristic histological appearance
consider: benign/uncertain neoplasm (when histology is not available) Y79

Y78 Malignant neoplasm genital, other, male ***C50, C60, C62, C63***

incl: testis, seminoma, carcinoma male breast
criteria: characteristic histological appearance
consider: benign/uncertain neoplasm (when histology is not available) Y79

Y79 Benign/uncertain neoplasm male genital

D07.4 to D07.6, D24, D29, D40, D48.6

incl: male breast
excl: prostatic hypertrophy Y85

Y80 Injury genital, male

S30.2, S31.2, S31.3, S31.5, S38.0, S38.2, T28.3, T28.8

incl: circumcision

Y81 Phimosis/redundant prepuce

N47

incl: paraphimosis
criteria: for redundant prepuce: excessive length of prepuce, with inability to retract over the glans penis; for phimosis: tightness of the prepuce which prevents retraction over the glans penis

Y82 Hypospadias

Q54

Y83 Undescended testicle

Q53

incl: cryptorchidism
excl: retractile testis Y84
criteria: the testicle has never been observed in the scrotum, and the testicle cannot be manipulated into the scrotum

Y84 Congenital anomaly male genital, other

Q55, Q56, Q83

incl: hermaphroditism, retractile testis

Y85 Benign prostatic hypertrophy

N40

incl: fibroma; hyperplasia; median bar of prostate; prostatic obstruction; prostatomegaly
criteria: enlarged, smooth, firm prostate demonstrated by palpation, cystoscopy, or imaging, with no evidence of prostatic carcinoma
consider: symptoms, complaints about urination U01–U05; retention of urine U08

Y86 Hydrocoele

N43.0 to N43.3

criteria: non-tender fluctuant swelling surrounding the testis or spermatic cord with transillumination of the swelling or imaging evidence
consider: symptom/complaint of scrotum or testis U05

Y99 Genital disease, other

A55, A57, A58, A63.8, N42.0 to N42.2, N42.8, N42.9, N43.4, N44, N48.0, N48.2, N48.5, N48.6, N48.8, N49.1, N49.2, N49.8, N49.9, N50.0, N50.1, N50.8, N50.9, N64.8, Z90.7

incl: disease of male breast; epididymal cyst; spermatocoele; torsion of the testis; other disease of male genitalia NOS

excl: sexually transmitted disease NOS A78; gynaecomastia Y16; carcinoma male breast Y78

Z Social problems

Component 1—Symptoms and complaints

Z01 Poverty/financial problem *Z59.5, Z59.6, Z59.7 to Z59.9*

Note: Problems with living conditions essentially require the patient's expression of concern about them, with agreement about the existence of the problem and desire for help. Whatever the objective living conditions patients can consider these as a problem. Labelling these problems requires acknowledgement of absolute differences in living conditions, as well as the individual's perception.

Z02 Food and water problem *Z58.6, Z59.4*

Note: Problems with living conditions essentially require the patient's expression of concern about them, with agreement about the existence of the problem and desire for help. Whatever the objective living conditions patients can consider these as a problem. Labelling these problems requires acknowledgement of absolute differences in living conditions, as well as the individual's perception.

Z03 Housing/neighbourhood problem *Z59.0 to Z59.3, Z59.8, Z59.9*

Note: Problems with living conditions essentially require the patient's expression of concern about them, with agreement about the existence of the problem and desire for help. Whatever the objective living conditions patients can consider these as a problem. Labelling these problems requires acknowledgement of absolute differences in living conditions, as well as the individual's perception.

Z04 Social cultural problem *Z60.1 to Z60.9*

incl: illegitimate pregnancy

excl: unwanted pregnancy W79

Note: Problems with living conditions essentially require the patient's expression of concern about them, with agreement about the existence of the problem and desire for help. Whatever the objective living conditions patients can consider these as a problem.

(*cont.*)

Labelling these problems requires acknowledgement of absolute differences in living conditions, as well as the individual's perception.

Z05 Work problem *Z56.1 to Z56.7, Z57*

Note: Problems with working conditions essentially require the patient's expression of concern about them, with agreement about the existence of the problem and desire for help. Whatever the objective working conditions patients can consider these as a problem. Labelling these problems requires acknowledgement of absolute differences in working conditions, as well as the individual's perception.

Z06 Unemployment problem *Z56.0*

excl: retirement problems P25
Note: Problems with unemployment essentially require the patient's expression of concern about them, with agreement about the existence of the problem and desire for help. Whatever the objective nature of the unemployment patients can consider this as a problem. Labelling these problems requires acknowledgement of absolute differences in unemployment, as well as the individual's perception.

Z07 Education problem *Z55*

incl: illiteracy
Note: Problems with education essentially require the patient's expression of concern about them, with agreement about the existence of the problem and desire for help. Whatever the objective education status patients can consider this as a problem. Labelling these problems requires acknowledgement of absolute differences in education, as well as the individual's perception.

Z08 Social welfare problem *Z59.7*

Note: Problems with social welfare essentially require the patient's expression of concern about them, with agreement about the existence of the problem and desire for help. Whatever the objective social welfare situation patients can consider this as a problem. Labelling these problems requires acknowledgement of absolute differences in social welfare, as well as the individual's perception.

Z09 Legal problem *Z65.0 to Z65.3*

Note: Problems with legal issues essentially require the patient's expression of concern about them, with agreement about the existence of the problem and desire for help. Whatever the objective legal issues patients can consider these as a problem. Labelling these problems requires acknowledgement of absolute differences in legal issues as well as the individual's perception.

Z10 Health care system problem *Z64.4, Z75*

Note: Problems with the health care system essentially require the patient's expression of concern about them, with agreement about the existence of the problem and desire

(*cont.*)

for help. Whatever the objective health care system patients can consider this as a problem. Labelling these problems requires acknowledgement of absolute differences in the health care system as well as the individual's perception.

Z11 Compliance/being ill problem *Z75*

incl: poor compliance
Note: The diagnosis of social problems arising due to being ill requires the patient's agreement on the existence of the problem and desire for help.

Z12 Relationship problem, partners *T74.0, T74.3, Z63.0*

incl: emotional abuse
excl: physical abuse Z25
Note: The diagnosis of problems in the relationship between family partners requires the patient's agreement on the existence of the problem and desire for help.

Z13 Partner behaviour problem *Z63.0*

incl: infidelity, physical abuse
Note: The diagnosis of problems arising from the behaviour of a family partner requires the patient's agreement on the existence of the problem and desire for help.

Z14 Partner illness problem *Z63.6*

Note: The diagnosis of problems arising from one or both family partners being ill requires the patient's agreement on the existence of the problem and desire for help.

Z15 Loss or death of partner *Z63.4, Z63.5*

incl: bereavement, divorce, separation
Note: The diagnosis of problems arising from the loss or death of a partner requires the patient's agreement on the existence of the problem and desire for help.

Z16 Relationship problem with child *T74.0, T74.3, Z61, Z62, Z63.8*

incl: child abuse (emotional)
excl: physical abuse Z25
Note: The diagnosis of problems in the relationship between adult and child; or between child and child requires the patient's agreement on the existence of the problem and desire for help.

Z18 Illness problem with child *Z63.6*

The diagnosis of social problems arising due to a child being ill requires the patient's agreement on the existence of the problem and desire for help.

Z19 Loss or death of child *Z63.4*

Note: The diagnosis of problems arising from the loss or death of a child in the family requires the patient's agreement on the existence of the problem and desire for help.

Z20 Relationship problem, parent/family *T74.0, Z63.1, Z63.8*

incl: parent, adult sibling(s), other family member
excl: partner Z12, child Z16, friend Z24
Note: The diagnosis of problems in the relationship between family members requires the patient's agreement on the existence of the problem and desire for help.

Z21 Behaviour problem, parent/family *Z63.1, Z63.9*

excl: behaviour of child P22, adolescent P23, partner Z13
Note: The diagnosis of problems arising from the behaviour of a family member requires the patient's agreement on the existence of the problem and desire for help.

Z22 Illness problem, parent/family *Z63.6, Z63.7*

excl: partner being ill Z14
The diagnosis of problems arising from the illness of a family member requires the patient's agreement on the existence of the problem and desire for help.

Z23 Loss/death of parent/family member *Z63.4*

excl: loss of child Z19, loss of partner Z15
Note: The diagnosis of problems arising from the loss or death of a family member requires the patient's agreement on the existence of the problem and desire for help.

Z24 Relationship problem, friends *Z63.9*

excl: family members Z20
Note: The diagnosis of problems in the relationship between friends requires the patient's agreement on the existence of the problem and desire for help.

Z25 Assault/harmful event *T74.1, T74.2, T74.8, T74.9, Z65.4, Z65.5*

incl: physical abuse of child or partner, rape, sexual attack
excl: physical problems see appropriate rubric(s) in other chapters; psychological problems see Chapter P
Note: The diagnosis of social problems arising from assaults and other harmful events requires the patient's agreement on the existence of the problem and desire for help.

Z27 Fear of a social problem *Z71.1*

incl: concern about having a social problem

(*cont.*)

excl: if patient has a social problem code the problem
criteria: concern or fear of a social problem in a patient without the problem or until the problem is confirmed

Z28 Social handicap ***Z73.4, Z73.6***

incl: social isolation, living alone
Note: The COOP/WONCA Charts are suitable for documenting the patient's functional status (see Chapter 8).

Z29 Social problem NOS

Z58.0 to Z58.5, Z58.8, Z58.9, Z63.2, Z63.3, Z63.7, Z63.8, Z63.9, Z64.0, Z64.1, Z65.8, Z65.9, Z72.6, Z72.8, Z72.9, Z73.2, Z73.5, Z73.8, Z73.9, Z76.1

incl: environmental problems, malingering, pollution

11 Conversion codes from ICD-10

The relationship between ICPC and ICD-10 is complex. There are some concepts in both which are not exactly represented in the other.[18] However, for most rubrics in each classification one or more corresponding rubrics in the other can be mapped. This has been done in both directions in this book.

Because of these complexities, quite often conversion of a code from one classification to the other and then re-conversion back again will not necessarily lead back to the same original code, because in each direction there may be several codes to choose from. Exact choices can only be made if the title of the condition is used with the help of a thesaurus. The point of having the code conversions in this book is only to indicate where the contents of rubrics in each classification overlap.

ICPC-2 to ICD-10

In the tabular list of ICPC-2 rubrics (Chapter 10) each rubric includes all the ICD-10 rubrics to which it relates. Where it relates to all of the three-digit ICD-10 code this is given; where it relates to only part of the three-digit ICD-10 code, all the relevant four-digit ICD-10 codes are given. However this does not imply that the ICD-10 codes listed relate only to the ICPC-2 rubric, since some ICD-10 rubrics relate to more than one ICPC-2 rubric, as can be seen from perusing the list of conversion codes from ICD-10 to ICPC-2 in this chapter.

ICD-10 to ICPC-2

In this chapter conversion codes are listed for all ICD-10 three-digit codes, and where not all of a three-digit code maps to the same ICPC-2 code, the conversions for all the four-digit ICD-10 codes are given. However the rubrics which are asterisked in ICD-10 to denote manifestations of other diseases are not included as they are never used for the main diagnosis; nor are rubrics in Chapter XX of ICD-10, External Causes of Morbidity and Mortality, as ICPC is in general not based on aetiology. Exclusion of these ICD-10 rubrics helps avoid confusion, and emphasizes that these conversions are indicative rather than exact or comprehensive.

* means that an exact conversion cannot be made

ICD-10	ICPC-2
A00	D70
A01	D70
A02	D70
A03	D70
A04	D70
A05	D70
A06	D70
A07	D70

A08	D70
A09	D73
A15	A70
A16	A70
A17	A70
A18	A70
A19	A70
A20	A78
A21	A78
A22	A78
A23	A78
A24	A78
A25	A78
A26	A78
A27	A78
A28	A78
A30	A78
A31	A78
A32.1	N71
A32.0	S76
A32.7	A78
A32.8	A78
A32.9	A78
A33	N72
A34	N72
A35	N72
A36	R83
A37	R71
A38	A78
A39.0	N71
A39.1	A78
A39.2	A78
A39.3	A78
A39.8	A78
A39.9	A78
A40	A78
A41	A78
A42	A78
A43	A78
A44	A78
A46	S76
A48.0	A78
A48.1	R81
A48.2	A78
A48.3	A78
A48.4	A78
A48.8	A78
A49	A78
A50	X70
A50	Y70
A51	X70
A51	Y70
A52	X70
A52	Y70
A53	X70
A53	Y70
A54	X71
A54	Y71
A55	X99
A55	Y99
A56	X92
A57	X99
A57	Y99
A58	X99
A58	Y99
A59.0	X73
A59.0	Y73
A59.8	A78
A59.9	A78
A60	X90
A60	Y72
A63.0	X91
A63.0	Y76
A63.8	X99
A63.8	Y99
A64	A78
A65	X70
A65	Y70
A66	S76
A67	S76
A68	A78
A69.0	D83
A69.1	D83
A69.2	A78
A69.8	A78
A69.9	A78
A70	A78
A71	F86
A74.0	F70
A74.8	A78
A74.9	A78
A75	A78
A77	A78
A78	A78
A79	A78
A80	N70
A81	N73
A82	A77
A83	N71
A84	N71
A85.0	N70
A85.1	N71
A85.2	N71
A85.8	N71
A86	N71
A87	N71
A88.0	A76
A88.1	H82
A88.8	N73
A89	N73
A90	A77
A91	A77
A92	A77
A93	A77
A94	A77
A95	A77
A96	A77
A98	A77
A99	A77
B00.0	S71
B00.1	S71
B00.2	D83
B00.2	R74
B00.3	N71
B00.4	N71
B00.5	F73
B00.7	A77
B00.8	S71
B00.9	S71
B01	A72
B02	S70
B03	A76
B04	A76
B05	A71
B06	A74
B07	S03
B08.0	A76
B08.1	S95
B08.2	A76
B08.3	A76
B08.4	A76
B08.5	R74
B08.8	A76
B09	A76
B09	D72
B15	D72
B16	D72
B17	D72
B18	D72
B19	D72
B20	B90
B21	B90
B22	B90
B23	B90
B24	B90
B25	A77
B26	D71
B27	A75
B30	F70
B33.0	A77
B33.1	A77
B33.2	K70
B33.3	A77
B33.8	A77
B34	A77
B35	S74
B36	S74
B37.0	D83
B37.1	R83
B37.2	S75
B37.3	X72
B37.4	U72
B37.4	X72
B37.4	Y75
B37.5	N71
B37.6	K70
B37.7	A78
B37.8	A78
B37.9	A78
B38	A78
B39	A78
B40	A78
B41	A78
B42	A78

B43	A78
B44	R83
B45	A78
B46	A78
B47	A78
B48	A78
B49	A78
B50	A73
B51	A73
B52	A73
B53	A73
B54	A73
B55	A78
B56	A78
B57	A78
B58.0	F73
B58.1	D97
B58.2	N71
B58.3	R83
B58.8	A78
B58.9	A78
B59	A78
B60	A78
B64	A78
B65	D96
B66	D96
B67	D96
B68	D96
B69	D96
B70	D96
B71	D96
B72	D96
B73	D96
B74	D96
B75	D96
B76	D96
B77	D96
B78	D96
B79	D96
B80	D96
B81	D96
B82	D96
B83	D96
B85	S73
B86	S72
B87	S73
B88.0	S72
B88.1	S73
B88.2	S72
B88.3	S73
B88.8	S73
B88.9	S73
B89	A78
B90	A70
B91	N70
B92	A78
B94.0	F86
B94.1	N71
B94.2	D97
B94.8	A78
B94.9	A78
B95	A78
B96	A78
B97	A77
B99	A78
C00	D77
C01	D77
C02	D77
C03	D77
C04	D77
C05	D77
C06	D77
C07	D77
C08	D77
C09	R85
C10	R85
C11	R85
C12	R85
C13	R85
C14.0	R85
C14.1	R85
C14.2	R85
C14.8	D77
C15	D77
C16	D74
C17	D77
C18	D75
C19	D75
C20	D75
C21	D75
C22	D77
C23	D77
C24	D77
C25	D76
C26	D77
C30.0	R85
C30.1	H75
C31	R85
C32	R85
C33	R84
C34	R84
C37	B74
C38.0	K72
C38.1	A79
C38.2	A79
C38.3	A79
C38.4	R85
C38.8	A79
C39	R85
C40	L71
C41	L71
C43	S77
C44	S77
C45.0	R85
C45.1	D77
C45.2	K72
C45.7	A79
C45.9	A79
C46.0	S77
C46.1	S77
C46.2	D77
C46.3	B74
C46.7	A79
C46.8	A79
C46.9	A79
C47	N74
C48	D77
C49	L71
C50	X76
C50	Y78
C51	X77
C52	X77
C53	X75
C54	X77
C55	X77
C56	X77
C57	X77
C58	W72
C60	Y78
C61	Y77
C62	Y78
C63	Y78
C64	U75
C65	U75
C66	U77
C67	U76
C68	U77
C69	F74
C70	N74
C71	N74
C72	N74
C73	T71
C74	T73
C75	T73
C76	A79
C77	B74
C78	A79
C79	A79
C80	A79
C81	B72
C82	B72
C83	B72
C84	B72
C85	B72
C88	B74
C90	B74
C91	B73
C92	B73
C93	B73
C94	B73
C95	B73
C96	B74
C97	A79
D00	D78
D01	D78
D02	R92
D03	S79
D04	S79
D05	X81
D06	X75
D07.0	X81
D07.1	X81
D07.2	X81
D07.3	X81

D07.4	Y79	*D39*	X81	*D65*	B83	*E26*	T99
D07.5	Y79	*D40*	Y79	*D66*	B83	*E27*	T99
D07.6	Y79	*D41*	U79	*D67*	B83	*E28*	T99
D09.0	U79	*D42*	N76	*D68*	B83	*E29*	T99
D09.1	U79	*D43*	N76	*D69*	B83	*E30*	T99
D09.2	F74	*D44*	T73	*D70*	B84	*E31*	T99
D09.3	T73	*D45*	B75	*D71*	B84	*E32*	T99
D09.7	A79	*D46*	B82	*D72*	B84	*E34.0*	T99
D09.9	A79	*D47*	B75	*D73*	B99	*E34.1*	T99
D10	D78	*D48.0*	L97	*D74*	B99	*E34.2*	T99
D11	D78	*D48.1*	L97	*D75*	B99	*E34.3*	T10
D12	D78	*D48.2*	N76	*D76*	B99	*E34.4*	T99
D13	D78	*D48.3*	D78	*D80*	B99	*E34.5*	T99
D14	R86	*D48.4*	D78	*D81*	B99	*E34.8*	T99
D15.0	B75	*D48.5*	S79	*D82*	B99	*E34.9*	T99
D15.1	K72	*D48.6*	X81	*D83*	B99	*E40*	T91
D15.2	K72	*D48.6*	Y79	*D84*	B99	*E41*	T91
D15.7	A99	*D48.7*	F74	*D86*	B99	*E42*	T91
D15.9	A99	*D48.7*	K72	*D89*	B99	*E43*	T91
D16	L97	*D48.8*	H75	*E00*	T80	*E44*	T91
D17	S78	*D48.9*	A99	*E01*	T86	*E45*	T91
D18	S81	*D50*	B80	*E02*	T86	*E46*	T91
D19	R86	*D51*	B81	*E03*	T86	*E50*	T91
D20	D78	*D52*	B81	*E04*	T81	*E51.1*	T91
D21	L97	*D53*	B82	*E05*	T85	*E51.2*	N99
D22	S82	*D55*	B82	*E06.0*	T70	*E51.8*	T91
D23	S79	*D56*	B78	*E06.1*	T99	*E51.9*	T91
D24	X79	*D57*	B78	*E06.2*	T99	*E52*	T91
D24	Y79	*D58*	B78	*E06.3*	T99	*E53*	T91
D25	X78	*D59*	B82	*E06.4*	T99	*E54*	T91
D26	X80	*D60*	B82	*E06.5*	T99	*E55*	T91
D27	X80	*D61.0*	B79	*E06.9*	T99	*E56*	T91
D28	X80	*D61.1*	A85	*E07*	T99	*E58*	T91
D29	Y79	*D61.2*	A86	*E10*	T89	*E59*	T91
D30	U78	*D61.3*	B82	*E11*	T90	*E60*	T91
D31	F74	*D61.8*	B82	*E12*	T90	*E61*	T91
D32	N75	*D61.9*	B82	*E13*	T90	*E63*	T91
D33	N75	*D62*	B82	*E14*	T90	*E64*	T91
D34	T72	*D64.0*	B79	*E15*	T87	*E65*	T99
D35	T73	*D64.1*	B82	*E16*	T87	*E66*	T82
D36.0	B75	*D64.2*	A85	*E20*	T99	*E66*	T83
D36.1	N75	*D64.2*	A86	*E21*	T99	*E67*	T99
D36.7	A99	*D64.3*	B82	*E22*	T99	*E68*	T99
D36.9	A99	*D64.4*	B79	*E23*	T99	*E70*	T99
D37	D78	*D64.8*	B82	*E24*	T99	*E71*	T99
D38	R92	*D64.9*	B82	*E25*	T99	*E72*	T99

E73	T99	*F22*	P72	*F50.1*	P86	*F88*	P99
E74	T99	*F23*	P98	*F50.2*	P86	*F89*	P99
E75	T99	*F24*	P72	*F50.3*	P86	*F90*	P81
E76	T99	*F25*	P72	*F50.4*	P02	*F91*	P22
E77	T99	*F28*	P72	*F50.4*	P86	*F91*	P23
E78	T93	*F29*	P72	*F50.5*	D10	*F92*	P22
E79	T99	*F30*	P73	*F50.8*	P29	*F92*	P23
E80	T99	*F31*	P73	*F50.9*	P29	*F93*	P22
E83	A91	*F32*	P76	*F51*	P06	*F94*	P22
E84	T99	*F33*	P76	*F52.0*	P07	*F94*	P23
E85	T99	*F34.0*	P73	*F52.1*	P08	*F95*	P10
E86	T11	*F34.1*	P76	*F52.2*	P08	*F98.0*	P12
E87	T99	*F34.8*	P76	*F52.3*	P08	*F98.1*	P13
E88	T99	*F34.9*	P76	*F52.4*	P08	*F98.2*	P11
E89	A87	*F38*	P76	*F52.5*	P08	*F98.3*	P11
F01	P70	*F39*	P76	*F52.8*	P08	*F98.4*	P10
F02	P70	*F40*	P79	*F52.9*	P08	*F98.5*	P10
F03	P70	*F41.0*	P74	*F53.0*	P76	*F98.6*	P10
F04	P71	*F41.1*	P74	*F53.1*	P98	*F98.8*	P22
F05	P71	*F41.2*	P76	*F53.8*	P99	*F98.8*	P23
F06	P71	*F41.3*	P74	*F53.9*	P99	*F98.8*	P29
F07	P71	*F41.4*	P74	*F54*	P99	*F98.9*	P22
F09	P71	*F41.5*	P74	*F55*	P18	*F98.9*	P23
F10.0	P16	*F41.6*	P74	*F59*	P99	*F98.9*	P29
F10.1	P15	*F41.7*	P74	*F60*	P80	*F99*	P77
F10.2	P15	*F41.8*	P74	*F61*	P80	*F99*	P99
F10.3	P15	*F41.9*	P01	*F62*	P80	*G00*	N71
F10.4	P15	*F41.9*	P74	*F63*	P80	*G03*	N71
F10.5	P15	*F42*	P79	*F63.3*	P29	*G04*	N71
F10.6	P15	*F43.0*	P02	*F64*	P09	*G06*	N73
F10.7	P15	*F43.1*	P82	*F65*	P09	*G08*	N73
F10.8	P15	*F43.2*	P02	*F66*	P09	*G09*	N73
F10.9	P15	*F43.8*	P02	*F68*	P80	*G10*	N99
F11	P19	*F43.9*	P02	*F69*	P80	*G11*	N99
F12	P19	*F44*	P75	*F70*	P85	*G12*	N99
F13	P18	*F45.1*	P75	*F71*	P85	*G20*	N87
F13	P19	*F45.2*	P75	*F72*	P85	*G21*	N87
F14	P19	*F45.3*	*	*F73*	P85	*G23*	N99
F15	P19	*F45.4*	*	*F78*	P85	*G24*	N99
F16	P19	*F45.8*	*	*F79*	P85	*G25.0*	N08
F17	P17	*F45.9*	*	*F80*	P24	*G25.1*	N08
F18	P19	*F48.0*	P78	*F80.9*	P22	*G25.2*	N08
F19	P18	*F48.1*	P79	*F81*	P24	*G25.3*	N08
F19	P19	*F48.8*	P79	*F82*	P24	*G25.4*	N08
F20	P72	*F48.9*	P79	*F83*	P24	*G25.5*	N08
F21	P72	*F50.0*	P86	*F84*	P99	*G25.8*	N04

G25.8	N08	*G81*	N99	*H05.4*	F99	*H35.8*	F99
G25.9	N08	*G82*	N99	*H05.5*	F99	*H35.9*	F99
G30	P70	*G83*	N99	*H05.8*	F99	*H40*	F93
G31.0	N99	*G90*	N99	*H05.9*	F99	*H43*	F99
G31.1	N99	*G91*	N99	*H10.0*	F70	*H44.0*	F99
G31.2	P15	*G92*	N99	*H10.1*	F71	*H44.1*	F99
G31.8	N99	*G93.0*	N99	*H10.2*	F70	*H44.2*	F99
G31.9	N99	*G93.1*	N99	*H10.3*	F70	*H44.3*	F99
G35	N86	*G93.2*	N99	*H10.4*	F70	*H44.4*	F99
G36	N99	*G93.3*	A04	*H10.5*	F70	*H44.5*	F99
G37	N99	*G93.4*	N99	*H10.8*	F70	*H44.6*	F79
G40	N88	*G93.5*	N99	*H10.9*	F70	*H44.7*	F79
G41	N88	*G93.6*	N99	*H11.0*	F99	*H44.8*	F99
G43	N89	*G93.7*	N99	*H11.1*	F99	*H44.9*	F99
G44.0	N90	*G93.8*	N99	*H11.2*	F99	*H46*	F99
G44.1	N89	*G93.9*	N99	*H11.3*	F75	*H47*	F99
G44.2	N95	*G95*	N99	*H11.4*	F99	*H49*	F95
G44.3	N01	*G96*	N99	*H11.8*	F99	*H50*	F95
G44.4	A85	*G97*	A87	*H11.9*	F99	*H51*	F95
G44.8	N01	*G98*	N18	*H15*	F99	*H52*	F91
G45	K89	*G98*	N99	*H16.0*	F85	*H53.0*	F99
G47	P06	*H00*	F72	*H16.1*	F73	*H53.1*	F05
G50.0	N92	*H01*	F72	*H16.2*	F73	*H53.2*	F05
G50.1	N03	*H02.0*	F99	*H16.3*	F73	*H53.3*	F05
G50.8	N92	*H02.1*	F99	*H16.4*	F73	*H53.4*	F99
G50.9	N92	*H02.2*	F16	*H16.8*	F73	*H53.5*	F99
G51	N91	*H02.3*	F16	*H16.9*	F73	*H53.6*	F99
G52	N99	*H02.4*	F16	*H17*	F99	*H53.8*	F04
G54	N94	*H02.5*	F16	*H18*	F99	*H53.9*	F05
G56.0	N93	*H02.6*	F16	*H20*	F73	*H54.0*	F94
G56.1	N94	*H02.7*	F16	*H21*	F73	*H54.1*	F94
G56.2	N94	*H02.8*	F99	*H25*	F92	*H54.2*	F94
G56.3	N94	*H02.9*	F99	*H26*	F92	*H54.3*	F94
G56.4	N94	*H04.0*	F99	*H27*	F99	*H54.4*	F28
G56.8	N94	*H04.1*	F99	*H30*	F73	*H54.5*	F28
G56.9	N94	*H04.2*	F03	*H31*	F99	*H54.6*	F28
G57	N94	*H04.3*	F73	*H33*	F82	*H54.7*	F05
G58	N94	*H04.4*	F73	*H34*	F99	*H55*	F14
G60	N94	*H04.5*	F99	*H35.0*	F83	*H57.0*	F99
G61	N94	*H04.6*	F99	*H35.1*	F83	*H57.1*	F01
G62	N94	*H04.8*	F99	*H35.2*	F83	*H57.8*	F02
G64	N94	*H04.9*	F99	*H35.3*	F84	*H57.8*	F13
G70	N99	*H05.0*	F73	*H35.4*	F83	*H57.8*	F15
G71	N99	*H05.1*	F73	*H35.5*	F99	*H57.8*	F99
G72	N99	*H05.2*	F99	*H35.6*	F99	*H57.9*	F29
G80	N99	*H05.3*	F99	*H35.7*	F99	*H59*	F99

H60	H70	*H93.8*	H13	*I61*	K90	*I95.1*	K88
H61.0	H99	*H93.8*	H99	*I62*	K90	*I95.2*	A85
H61.1	H99	*H93.9*	H29	*I63*	K90	*I95.8*	K88
H61.2	H81	*H95*	A87	*I64*	K90	*I95.9*	K88
H61.3	H99	*I00*	K71	*I65*	K91	*I97*	A87
H61.8	H99	*I01*	K71	*I66*	K91	*I99*	K99
H61.9	H99	*I02*	K71	*I67.0*	K91	*J00*	R74
H65	H72	*I05*	K71	*I67.1*	K91	*J01*	R75
H66.0	H71	*I06*	K71	*I67.2*	K91	*J02.0*	R72
H66.1	H74	*I07*	K71	*I67.3*	K91	*J02.8*	R74
H66.2	H74	*I08*	K83	*I67.4*	K87	*J02.9*	R74
H66.3	H74	*I09*	K71	*I67.5*	K91	*J03.0*	R72
H66.4	H71	*I10*	K86	*I67.6*	K91	*J03.8*	R76
H66.9	H71	*I11*	K87	*I67.7*	K91	*J03.9*	R76
H68	H73	*I12*	K87	*I67.8*	K91	*J04*	R77
H69	H73	*I13*	K87	*I67.9*	K91	*J05.0*	R77
H70	H74	*I15*	K87	*I69*	K91	*J05.1*	R83
H71	H74	*I20*	K74	*I70*	K92	*J06*	R74
H72	H77	*I21*	K75	*I71*	K99	*J10.0*	R81
H73.0	H71	*I22*	K75	*I72*	K99	*J10.1*	R80
H73.1	H74	*I23*	K75	*I73*	K92	*J10.8*	R80
H73.8	H99	*I24*	K74	*I74*	K92	*J11.0*	R81
H73.9	H99	*I24*	K76	*I77*	K99	*J11.1*	R80
H74	H99	*I25*	K76	*I78.0*	K99	*J11.8*	R80
H80	H83	*I26*	K93	*I78.1*	K06	*J12*	R81
H81	H82	*I27*	K82	*I78.8*	K99	*J13*	R81
H83.0	H82	*I28*	K82	*I78.9*	K99	*J14*	R81
H83.1	H99	*I30*	K70	*I80*	K94	*J15*	R81
H83.2	H99	*I31*	K84	*I81*	K94	*J16*	R81
H83.3	H85	*I33*	K70	*I82*	K94	*J18*	R81
H83.8	H99	*I34*	K83	*I83.0*	S97	*J20*	R78
H83.9	H99	*I35*	K83	*I83.1*	K95	*J21*	R78
H90	H86	*I36*	K83	*I83.2*	S97	*J22*	R78
H91.0	H86	*I37*	K83	*I83.9*	K95	*J30*	R97
H91.1	H84	*I38*	K70	*I84*	K96	*J31*	R83
H91.2	H86	*I40*	K70	*I85*	K99	*J32*	R75
H91.3	H86	*I42*	K84	*I86*	K99	*J33*	R99
H91.8	H86	*I44*	K84	*I87.0*	K94	*J34.0*	R73
H91.9	H86	*I45*	K84	*I87.1*	K99	*J34.1*	R99
H92.0	H01	*I46*	K84	*I87.2*	K95	*J34.2*	R99
H92.1	H04	*I47*	K79	*I87.8*	K06	*J34.3*	R99
H92.2	H05	*I48*	K78	*I87.8*	K95	*J34.8*	R08
H93.0	H99	*I49*	K80	*I87.9*	K99	*J34.8*	R09
H93.1	H03	*I50*	K77	*I88*	B71	*J34.8*	R99
H93.2	H02	*I51*	K84	*I89*	B99	*J35*	R90
H93.3	H99	*I60*	K90	*I95.0*	K88	*J36*	R76

J37	R83	*K02*	D82	*K56.2*	D99	*K91.2*	D99
J38	R99	*K03*	D82	*K56.3*	D99	*K91.3*	A87
J39	R99	*K04*	D82	*K56.4*	D12	*K91.4*	A89
J40	R78	*K05*	D82	*K56.5*	D99	*K91.5*	D99
J41	R79	*K06*	D82	*K56.6*	D99	*K91.8*	D99
J42	R79	*K07*	D82	*K56.7*	D99	*K91.9*	D99
J43	R95	*K08*	D82	*K57*	D92	*K92.0*	D14
J44	R95	*K09*	D82	*K58*	D93	*K92.1*	D15
J45	R96	*K10*	D82	*K59.0*	D12	*K92.2*	D99
J46	R96	*K11*	D83	*K59.1*	D11	*K92.8*	D99
J47	R99	*K12*	D83	*K59.2*	D99	*K92.9*	D99
J60	R99	*K13*	D83	*K59.3*	D99	*L00*	S84
J61	R99	*K14*	D20	*K59.4*	D04	*L01*	S84
J62	R99	*K14*	D83	*K59.8*	D99	*L02*	S10
J63	R99	*K20*	D84	*K59.9*	D99	*L03.0*	S09
J64	R99	*K21*	D84	*K60*	D95	*L03.1*	S76
J65	R99	*K22*	D84	*K61*	D95	*L03.2*	S76
J66	R99	*K25*	D86	*K62.0*	D78	*L03.3*	S76
J67	R99	*K26*	D85	*K62.1*	D78	*L03.8*	S76
J68	R99	*K27*	D86	*K62.2*	D99	*L03.9*	S76
J69	R99	*K28*	D86	*K62.3*	D99	*L04*	B70
J70	R99	*K29*	D87	*K62.4*	D99	*L05*	S85
J80	R99	*K30*	D07	*K62.5*	D16	*L08*	S76
J81	R99	*K31*	D99	*K62.6*	D99	*L10*	S99
J82	R99	*K35*	D88	*K62.7*	D99	*L11*	S99
J84	R99	*K36*	D88	*K62.8*	D99	*L12*	S99
J85	R83	*K37*	D88	*K62.9*	D99	*L13*	S99
J86	R83	*K38*	D99	*K63*	D99	*L20*	S87
J90	R82	*K40*	D89	*K65*	D99	*L21*	S86
J92	R99	*K41*	D91	*K66*	D99	*L22*	S89
J93	R99	*K42*	D91	*K70*	D97	*L23*	S88
J94	R82	*K43*	D91	*K71*	D97	*L24*	S88
J95	A87	*K44*	D90	*K72*	D97	*L25*	S88
J96	R99	*K45*	D91	*K73*	D97	*L26*	S99
J98	R99	*K46*	D91	*K74*	D97	*L27.0*	A85
K00.0	D82	*K50*	D94	*K75*	D97	*L27.1*	A85
K00.1	D82	*K51*	D94	*K76*	D97	*L27.2*	S88
K00.2	D82	*K52.0*	D94	*K80*	D98	*L27.8*	S88
K00.3	D82	*K52.1*	D99	*K81*	D98	*L27.9*	S88
K00.4	D82	*K52.2*	D99	*K82*	D98	*L28*	S99
K00.5	D82	*K52.8*	D99	*K83*	D98	*L29.0*	D05
K00.6	D82	*K52.9*	D11	*K85*	D99	*L29.1*	Y05
K00.7	D19	*K52.9*	D99	*K86*	D99	*L29.2*	X16
K00.8	D82	*K55*	D99	*K90*	D99	*L29.3*	D05
K00.9	D82	*K56.0*	D99	*K91.0*	A87	*L29.8*	S02
K01	D82	*K56.1*	D99	*K91.1*	D99	*L29.9*	S02

L30.0	S88	*L72.2*	S99	*M02*	L99	*M25.5*	L20
L30.1	S92	*L72.8*	S99	*M05*	L88	*M25.6*	L08
L30.2	S99	*L72.9*	S99	*M06*	L88	*M25.6*	L10
L30.3	S88	*L73.0*	S99	*M08*	L88	*M25.6*	L11
L30.4	S88	*L73.1*	S99	*M10*	T92	*M25.6*	L12
L30.5	S99	*L73.2*	S92	*M11*	T99	*M25.6*	L13
L30.8	S88	*L73.8*	S99	*M12*	L99	*M25.6*	L15
L30.9	S88	*L73.9*	S99	*M13*	L91	*M25.6*	L16
L40	S91	*L74*	S92	*M15*	L91	*M25.6*	L17
L41	S99	*L75*	S92	*M16*	L89	*M25.6*	L20
L42	S90	*L80*	S99	*M17*	L90	*M25.7*	L99
L43	S99	*L81.0*	S08	*M18*	L91	*M25.8*	L20
L44	S99	*L81.1*	S08	*M19*	L91	*M25.8*	L99
L50	S98	*L81.2*	S08	*M20*	L98	*M25.9*	L20
L51	S99	*L81.3*	S08	*M21*	L98	*M25.9*	L99
L52	S99	*L81.4*	S99	*M22.0*	L80	*M30*	K99
L53.0	S99	*L81.5*	S99	*M22.1*	L80	*M31*	K99
L53.1	S99	*L81.6*	S99	*M22.2*	L99	*M32*	L99
L53.2	S99	*L81.7*	S99	*M22.3*	L99	*M33*	L99
L53.3	S99	*L81.8*	S99	*M22.4*	L99	*M34*	L99
L53.8	S99	*L81.9*	S99	*M22.8*	L99	*M35*	L99
L53.9	S06	*L82*	S99	*M22.9*	L99	*M40*	L85
L53.9	S07	*L83*	S99	*M23*	L99	*M41*	L85
L55	S80	*L84*	S20	*M24*	L99	*M42*	L94
L56	S80	*L85*	S99	*M25.0*	L99	*M43.0*	L84
L57	S80	*L87*	S99	*M25.1*	L99	*M43.1*	L84
L58	S80	*L88*	S99	*M25.2*	L99	*M43.2*	L99
L59	S80	*L89*	S97	*M25.3*	L99	*M43.3*	L83
L60.0	S94	*L90*	S99	*M25.4*	L08	*M43.4*	L83
L60.1	S22	*L91*	S99	*M25.4*	L10	*M43.5*	L84
L60.2	S99	*L92*	S99	*M25.4*	L11	*M43.6*	L83
L60.3	S99	*L93*	S99	*M25.4*	L12	*M43.8*	L85
L60.4	S22	*L94*	S99	*M25.4*	L13	*M43.9*	L85
L60.5	S22	*L95*	S99	*M25.4*	L15	*M45*	L88
L60.8	S99	*L97*	S97	*M25.4*	L16	*M46.0*	L84
L60.9	S22	*L98.0*	S76	*M25.4*	L17	*M46.1*	L84
L63	S23	*L98.1*	S99	*M25.4*	L20	*M46.2*	L70
L64	S23	*L98.2*	S99	*M25.5*	L07	*M46.3*	L70
L65	S23	*L98.3*	S99	*M25.5*	L08	*M46.4*	L70
L66	S23	*L98.4*	S97	*M25.5*	L10	*M46.5*	L70
L67	S24	*L98.5*	S99	*M25.5*	L11	*M46.8*	L84
L68	S24	*L98.6*	S99	*M25.5*	L12	*M46.9*	L84
L70	S96	*L98.7*	S99	*M25.5*	L13	*M47.0*	L84
L71	S99	*L98.8*	S99	*M25.5*	L15	*M47.1*	L86
L72.0	S99	*L98.9*	S99	*M25.5*	L16	*M47.2*	L86
L72.1	S93	*M00*	L70	*M25.5*	L17	*M47.8*	L83

M47.8	L84
M47.9	L83
M47.9	L84
M48	L83
M48	L84
M50	L83
M51	L84
M51	L86
M53.0	L83
M53.1	L83
M53.2	L84
M53.3	L03
M53.3	L84
M53.8	L84
M53.9	L84
M54.0	L01
M54.0	L02
M54.0	L03
M54.1	L99
M54.2	L01
M54.3	L86
M54.4	L86
M54.5	L03
M54.6	L02
M54.8	L02
M54.9	L02
M60.0	L70
M60.1	L18
M60.2	L18
M60.8	L18
M60.9	L18
M61	L99
M62.0	L99
M62.1	L99
M62.2	L99
M62.3	L99
M62.4	L99
M62.5	L19
M62.6	L19
M62.8	L99
M62.9	L99
M65.0	L70
M65.1	L70
M65.2	L87
M65.3	L87
M65.4	L87

M65.8	L87
M65.9	L87
M66	L99
M67.0	L99
M67.1	L99
M67.2	L99
M67.3	L87
M67.4	L87
M67.8	L99
M67.9	L99
M70	L87
M71.0	L70
M71.1	L70
M71.2	L87
M71.3	L87
M71.4	L87
M71.5	L87
M71.8	L87
M71.9	L87
M72	L87
M75	L92
M76	L87
M77.0	L87
M77.1	L93
M77.2	L87
M77.3	L87
M77.4	L17
M77.5	L17
M77.8	L87
M77.9	L87
M79.0	L18
M79.1	L18
M79.2	N29
M79.2	N99
M79.3	L99
M79.4	L99
M79.5	L81
M79.6	L09
M79.6	L18
M79.8	L99
M79.9	L19
M80	L95
M81	L95
M83	T99
M84	L99
M85	L99

M86	L70
M87	L99
M88	L99
M89	L99
M91	L94
M92	L94
M93	L94
M94	L99
M95	L99
M96	A87
M99	L99
N00	U88
N01	U88
N02	U06
N03	U88
N04	U88
N05	U88
N06	U99
N07	U88
N10	U70
N11	U70
N12	U70
N13	U99
N14	U88
N15.0	U88
N15.1	U70
N15.8	U88
N15.9	U70
N17	U99
N18	U99
N19	U99
N20	U95
N21	U95
N23	U14
N25	U99
N26	U99
N27	U99
N28	U99
N30	U71
N31	U99
N32	U99
N34	U72
N35	U99
N36	U99
N39.0	U71
N39.1	U98

N39.2	U90
N39.3	U04
N39.4	U04
N39.8	U99
N39.9	U99
N40	Y06
N40	Y85
N41	Y73
N42.0	Y99
N42.1	Y99
N42.2	Y99
N42.8	Y99
N42.9	Y06
N42.9	Y99
N43.0	Y86
N43.1	Y86
N43.2	Y86
N43.3	Y86
N43.4	Y99
N44	Y99
N45	Y74
N46	Y10
N47	Y81
N48.0	Y99
N48.1	Y75
N48.2	Y99
N48.3	Y08
N48.4	Y07
N48.5	Y99
N48.6	Y99
N48.8	Y99
N48.9	Y01
N48.9	Y04
N48.9	Y08
N49.0	Y73
N49.1	Y99
N49.2	Y99
N49.8	Y99
N49.9	Y99
N50.0	Y99
N50.1	Y99
N50.8	Y99
N50.9	Y02
N50.9	Y05
N50.9	Y29
N50.9	Y99

N60	X88	*N89.9*	X15	*O02*	W82	*O46*	W03
N61	X21	*N90.0*	X99	*O03*	W82	*O47*	W99
N61	X99	*N90.1*	X99	*O04*	W83	*O48*	W99
N62	X21	*N90.2*	X99	*O05*	W82	*O60*	W92
N62	Y16	*N90.3*	X99	*O06*	W82	*O60*	W93
N63	X19	*N90.4*	X99	*O07*	W99	*O61*	W92
N63	Y16	*N90.5*	X99	*O08*	W99	*O61*	W93
N64.0	X20	*N90.6*	X99	*O10*	W81	*O62*	W92
N64.1	X99	*N90.7*	X99	*O11*	W81	*O62*	W93
N64.2	X99	*N90.8*	X99	*O12*	W81	*O63*	W92
N64.3	X21	*N90.9*	X16	*O13*	W81	*O63*	W93
N64.4	X18	*N91*	X05	*O14*	W81	*O64*	W92
N64.5	X20	*N92.0*	X06	*O15*	W81	*O64*	W93
N64.5	X21	*N92.1*	X07	*O16*	W81	*O65*	W92
N64.8	X99	*N92.2*	X06	*O20*	W03	*O65*	W93
N64.8	Y99	*N92.3*	X08	*O21*	W05	*O66*	W92
N64.9	X21	*N92.4*	X06	*O22*	W99	*O66*	W93
N70	X74	*N92.5*	X07	*O23*	W71	*O67*	W92
N71	X74	*N92.6*	X07	*O24.0*	W84	*O67*	W93
N72	X85	*N93.0*	X13	*O24.1*	W84	*O68*	W92
N73	X74	*N93.8*	X08	*O24.2*	W84	*O68*	W93
N75	X99	*N93.9*	X08	*O24.3*	W84	*O69*	W92
N76	X84	*N94.0*	X03	*O24.4*	W85	*O69*	W93
N80	X99	*N94.1*	X04	*O24.9*	W84	*O70*	W92
N81	X87	*N94.2*	X04	*O25*	W84	*O70*	W93
N82	X99	*N94.3*	X89	*O26*	W99	*O71*	W92
N83	X99	*N94.4*	X02	*O28*	W99	*O71*	W93
N84.0	X99	*N94.5*	X02	*O29*	A87	*O72*	W17
N84.1	X85	*N94.6*	X02	*O30*	W84	*O73*	W92
N84.2	X99	*N94.8*	X99	*O31*	W84	*O73*	W93
N84.3	X99	*N94.9*	X09	*O32*	W84	*O74*	A87
N84.8	X99	*N94.9*	X17	*O33*	W84	*O75*	W92
N84.9	X99	*N94.9*	X29	*O34*	W84	*O75.0*	W92
N85	X99	*N94.9*	X99	*O35*	W84	*O75.0*	W93
N86	X85	*N95.0*	X12	*O36*	W84	*O75.1*	W92
N87	X85	*N95.1*	X11	*O40*	W84	*O75.1*	W93
N88	X85	*N95.2*	X11	*O41.0*	W99	*O75.2*	W71
N89.0	X99	*N95.3*	X11	*O41.1*	W71	*O75.3*	W71
N89.1	X99	*N95.8*	X11	*O41.8*	W99	*O75.4*	W92
N89.2	X99	*N95.9*	X11	*O41.9*	W99	*O75.4*	W93
N89.3	X99	*N96*	X99	*O42*	W92	*O75.5*	W92
N89.4	X99	*N97*	W15	*O42*	W93	*O75.5*	W93
N89.5	X99	*N98*	X99	*O43*	W84	*O75.6*	W92
N89.6	X99	*N99*	A87	*O44*	W84	*O75.6*	W93
N89.7	X99	*O00*	W80	*O45*	W92	*O75.7*	W92
N89.8	X14	*O01*	W73	*O45*	W93	*O75.7*	W93

O75.8	W92	*O99*	W84	*P74*	A94	*Q28*	K73
O75.8	W93	*P00*	A94	*P76*	A94	*Q30*	R89
O75.9	W92	*P01*	A94	*P77*	A94	*Q31*	R89
O75.9	W93	*P02*	A94	*P78*	A94	*Q32*	R89
O80	W90	*P03*	A94	*P80*	A94	*Q33*	R89
O81	W92	*P04*	A94	*P81*	A94	*Q34*	R89
O81	W93	*P05*	A94	*P83*	A94	*Q35*	D81
O82	W92	*P07*	A93	*P90*	A94	*Q36*	D81
O82	W93	*P08*	A94	*P91*	A94	*Q37*	D81
O83	W92	*P10*	A94	*P92*	A94	*Q38*	D81
O83	W93	*P11*	A94	*P93*	A94	*Q39*	D81
O84	W92	*P12*	A94	*P94*	A94	*Q40*	D81
O84	W93	*P13*	A94	*P95*	A95	*Q41*	D81
O85	W70	*P14*	A94	*P96*	A94	*Q42*	D81
O86.0	A87	*P15*	A94	*Q00*	N85	*Q43*	D81
O86.1	W70	*P20*	A94	*Q01*	N85	*Q44*	D81
O86.2	W71	*P21*	A94	*Q02*	N85	*Q45*	D81
O86.3	W70	*P22*	A94	*Q03*	N85	*Q50*	X83
O86.4	W71	*P23*	A94	*Q04*	N85	*Q51*	X83
O86.8	W71	*P24*	A94	*Q05*	N85	*Q52*	X83
O87	W96	*P25*	A94	*Q06*	N85	*Q53*	Y83
O88	W99	*P26*	A94	*Q07*	N85	*Q54*	Y82
O89	A87	*P27*	A94	*Q10.0*	F81	*Q55*	Y84
O90.0	A87	*P28*	A94	*Q10.1*	F81	*Q56*	X83
O90.1	A87	*P29*	A94	*Q10.2*	F81	*Q56*	Y84
O90.2	A87	*P35*	A74	*Q10.3*	F81	*Q60*	U85
O90.3	K84	*P35*	A94	*Q10.4*	F81	*Q61*	U85
O90.4	W96	*P36*	A94	*Q10.5*	F80	*Q62*	U85
O90.5	W99	*P37*	A94	*Q10.6*	F81	*Q63*	U85
O90.8	W96	*P38*	A94	*Q10.7*	F81	*Q64*	U85
O90.9	W18	*P39*	A94	*Q11*	F81	*Q65*	L82
O90.9	W96	*P50*	A94	*Q12*	F81	*Q66*	L82
O91	W94	*P51*	A94	*Q13*	F81	*Q67*	L82
O92.0	W95	*P52*	A94	*Q14*	F81	*Q68*	L82
O92.1	W95	*P53*	A94	*Q15*	F81	*Q69*	L82
O92.2	W95	*P54*	A94	*Q16*	H80	*Q70*	L82
O92.3	W95	*P55*	A94	*Q17*	H80	*Q71*	L82
O92.4	W95	*P56*	A94	*Q18*	D81	*Q72*	L82
O92.5	W19	*P57*	A94	*Q20*	K73	*Q73*	L82
O92.6	W19	*P58*	A94	*Q21*	K73	*Q74*	L82
O92.7	W19	*P59*	A94	*Q22*	K73	*Q75*	L82
O95	W99	*P60*	A94	*Q23*	K73	*Q76*	L82
O96	W99	*P61*	A94	*Q24*	K73	*Q77*	L82
O97	W99	*P70*	A94	*Q25*	K73	*Q78*	L82
O98	W71	*P71*	A94	*Q26*	K73	*Q79*	L82
O99	W76	*P72*	A94	*Q27*	K73	*Q80*	S83

Q81	S83
Q82	S83
Q83	X83
Q83	Y84
Q84	S83
Q85	A90
Q86	A90
Q87	A90
Q89.0	B79
Q89.1	T80
Q89.2	T78
Q89.2	T80
Q89.3	A90
Q89.4	A90
Q89.7	A90
Q89.8	B79
Q89.9	A90
Q90	A90
Q91	A90
Q92	A90
Q93	A90
Q95	A90
Q96	A90
Q97	A90
Q98	A90
Q99	A90
R00.0	K04
R00.1	K04
R00.2	K04
R00.8	K05
R01	K81
R02	K92
R03.0	K85
R03.1	K29
R04.0	R06
R04.1	R29
R04.2	R24
R04.8	R29
R04.9	R29
R05	R05
R06.0	R02
R06.1	R04
R06.2	R03
R06.3	R04
R06.4	R98
R06.5	R04
R06.6	R29
R06.7	R07
R06.8	R04
R07.0	R21
R07.1	R01
R07.2	K01
R07.2	K02
R07.3	A11
R07.4	A11
R09.0	R29
R09.1	R82
R09.2	R29
R09.3	R25
R09.8	K03
R09.8	K29
R09.8	R21
R09.8	R29
R10.0	D01
R10.1	D02
R10.1	D06
R10.2	D04
R10.2	D06
R10.2	Y02
R10.3	D04
R10.3	D06
R10.4	D01
R11	D09
R11	D10
R12	D03
R13	D21
R14	D08
R15	D17
R16.0	D23
R16.1	B87
R16.2	B87
R16.2	D23
R17	D13
R18	D29
R19.0	D24
R19.0	D25
R19.1	D29
R19.2	D29
R19.3	D29
R19.4	D18
R19.5	D18
R19.6	D20
R19.8	D29
R20.0	N06
R20.1	N06
R20.2	N05
R20.3	N06
R20.8	N06
R20.8	S01
R21	S05
R21	S06
R21	S07
R22	S04
R23.0	S08
R23.1	S08
R23.2	S08
R23.3	S29
R23.4	S21
R23.8	S29
R25.0	N08
R25.1	N08
R25.2	L14
R25.3	N08
R25.8	N08
R26	N29
R27	N29
R29.0	N08
R29.0	N29
R29.1	N29
R29.1	N29
R29.2	N29
R29.3	L29
R29.4	L13
R29.8	L04
R29.8	L05
R29.8	L07
R29.8	L09
R29.8	L12
R29.8	L14
R29.8	L17
R29.8	L29
R29.8	N29
R30	U01
R31	U06
R32	U04
R33	U08
R34	U05
R35	U02
R36	X29
R36	Y03
R39.0	U13
R39.1	U05
R39.2	U99
R39.8	U07
R39.8	U13
R40	A07
R41	P20
R42	N17
R43	N16
R44	P29
R45.0	P01
R45.1	P04
R45.2	P03
R45.3	P03
R45.4	P04
R45.5	P04
R45.6	P04
R45.7	P29
R45.8	P29
R46	P29
R46.8	A18
R46.8	H15
R46.8	W21
R46.8	X22
R47	N19
R48	P24
R49	R23
R50	A02
R50	A03
R51	N01
R51	N03
R52	A01
R52	S01
R52	X01
R52	Y01
R52	Y02
R53	A04
R53	A05
R54	P05
R55	A06
R56	N07
R57	K99
R58	A10
R59	B02

R60	K07	*R90*	A91	*S02.5*	D80	*S12*	L76
R61	A09	*R91*	A91	*S02.6*	L76	*S13.0*	L80
R62.0	P22	*R92*	A91	*S02.7*	L76	*S13.1*	L80
R62.8	T10	*R93*	A91	*S02.8*	L76	*S13.2*	L80
R62.9	T10	*R94*	A91	*S02.9*	L76	*S13.3*	L80
R63.0	T03	*R95*	A95	*S03.0*	L80	*S13.4*	L83
R63.1	T01	*R95*	A96	*S03.1*	R88	*S13.5*	L83
R63.2	T02	*R96*	A96	*S03.2*	D80	*S13.6*	L83
R63.3	T04	*R98*	A96	*S03.3*	L80	*S14*	N81
R63.3	T05	*R99*	A96	*S03.4*	L79	*S15*	A80
R63.4	T08	*S00.0*	S16	*S03.5*	L79	*S16*	L81
R63.5	T07	*S00.0*	S17	*S04*	N81	*S17*	L81
R63.8	T29	*S00.0*	S18	*S05.0*	F79	*S18*	L81
R64	T08	*S00.0*	S19	*S05.1*	F75	*S19*	L81
R68.0	A29	*S00.1*	F75	*S05.2*	F79	*S20*	S16
R68.1	A16	*S00.2*	F79	*S05.3*	F79	*S20*	S17
R68.2	D20	*S00.3*	R88	*S05.4*	F79	*S20*	S19
R68.3	S22	*S00.4*	H78	*S05.5*	F79	*S21*	A80
R68.8	A02	*S00.5*	D80	*S05.6*	F79	*S21*	S18
R68.8	A08	*S00.7*	S16	*S05.7*	F79	*S22*	L76
R68.8	A29	*S00.7*	S17	*S05.8*	F79	*S23.0*	L81
R68.8	B04	*S00.7*	S18	*S05.9*	F79	*S23.1*	L80
R68.8	B29	*S00.7*	S19	*S06.0*	N79	*S23.2*	L80
R68.8	W29	*S00.8*	S16	*S06.1*	N80	*S23.3*	L79
R69	A99	*S00.8*	S17	*S06.2*	N80	*S23.4*	L79
R70	B99	*S00.8*	S18	*S06.3*	N80	*S23.5*	L79
R71	B99	*S00.8*	S19	*S06.4*	N80	*S24*	N81
R72	B84	*S00.9*	S16	*S06.5*	N80	*S25*	A80
R73	A91	*S00.9*	S17	*S06.6*	N80	*S26*	A80
R74	A91	*S00.9*	S18	*S06.7*	N80	*S27*	A80
R75	B90	*S00.9*	S19	*S06.8*	N80	*S27*	R88
R76	A91	*S00.9*	H78	*S06.9*	N80	*S28*	L81
R77	A91	*S01.0*	S18	*S07*	N80	*S29*	L81
R78	A91	*S01.1*	F79	*S08*	N80	*S30.0*	S16
R79	A91	*S01.2*	S18	*S09.0*	N80	*S30.1*	S16
R80	U98	*S01.3*	H79	*S09.1*	L81	*S30.2*	X82
R81	U98	*S01.4*	S18	*S09.2*	H79	*S30.2*	Y80
R82	U98	*S01.5*	D80	*S09.7*	N80	*S30.7*	S17
R83	A91	*S01.7*	S18	*S09.8*	N80	*S30.7*	S19
R84	A91	*S01.8*	S18	*S09.9*	H78	*S30.8*	S17
R85	A91	*S01.9*	S18	*S09.9*	N81	*S30.8*	S19
R86	A91	*S02.0*	N80	*S10*	S16	*S30.9*	S17
R87	A91	*S02.1*	N80	*S10*	S17	*S30.9*	S19
R87	X86	*S02.2*	L76	*S10*	S19	*S31.0*	A80
R88	A91	*S02.3*	L76	*S11*	A80	*S31.0*	S18
R89	A91	*S02.4*	L76	*S11*	S18	*S31.1*	A81

S31.1	S18
S31.2	Y80
S31.3	Y80
S31.4	X82
S31.5	X82
S31.5	Y80
S31.7	A81
S31.8	S18
S32	L76
S33.0	L81
S33.1	L80
S33.2	L80
S33.3	L80
S33.4	L81
S33.5	L84
S33.6	L79
S33.7	L84
S34	N81
S35	A80
S36.0	B76
S36.1	D80
S36.2	D80
S36.3	D80
S36.4	D80
S36.5	D80
S36.6	D80
S36.7	A81
S36.8	A81
S36.9	A81
S37.0	U80
S37.1	U80
S37.2	U80
S37.3	U80
S37.4	X82
S37.5	X82
S37.6	X82
S37.7	A81
S37.8	A81
S37.9	A80
S38.0	X82
S38.0	Y80
S38.1	A80
S38.2	X82
S38.2	Y80
S38.3	A80
S39	A80
S39	A81
S39	L81
S40	S16
S40	S17
S40	S19
S41	S18
S42	L76
S43.0	L80
S43.1	L80
S43.2	L80
S43.3	L80
S43.4	L79
S43.5	L79
S43.6	L79
S43.7	L79
S44	N81
S45	A80
S46	L81
S47	L81
S48	L81
S49	L81
S50	S16
S50	S17
S50	S19
S51	S18
S52	L72
S53.0	L80
S53.1	L80
S53.2	L79
S53.3	L79
S53.4	L79
S54	N81
S55	A80
S56	L81
S57	L81
S58	L81
S59	L81
S60	S16
S60	S17
S60	S19
S61	S18
S62	L74
S63.0	L80
S63.1	L80
S63.2	L80
S63.3	L79
S63.4	L79
S63.5	L79
S63.6	L79
S63.7	L79
S64	N81
S65	A80
S66	L81
S67	L81
S68	L81
S69	L81
S70	S16
S70	S17
S70	S19
S71	S18
S72	L75
S73.0	L80
S73.1	L79
S74	N81
S75	A80
S76	L81
S77	L81
S78	L81
S79	L81
S80	S16
S80	S17
S80	S19
S81	S18
S82.0	L76
S82.1	L73
S82.2	L73
S82.3	L73
S82.4	L73
S82.5	L73
S82.6	L73
S82.7	L73
S82.8	L73
S82.9	L73
S83.0	L80
S83.1	L80
S83.2	L96
S83.3	L96
S83.4	L78
S83.5	L96
S83.6	L78
S83.7	L96
S84	N81
S85	A80
S86	L81
S87	L81
S88	L81
S89	L81
S90	S16
S90	S17
S90	S19
S91	S18
S92	L74
S93.0	L80
S93.1	L80
S93.2	L79
S93.3	L80
S93.4	L77
S93.5	L79
S93.6	L79
S94	N81
S95	A80
S96	L81
S97	L81
S98	L81
S99	L81
T00	A81
T01	A81
T02	A81
T03	A81
T04	A81
T05	A81
T06.0	N81
T06.1	N81
T06.2	N81
T06.3	K99
T06.4	L81
T06.5	A81
T06.6	A81
T06.7	A81
T06.8	A81
T07	A81
T08	L76
T09	A80
T10	L76
T11	L81
T12	L76
T13	L81
T14.0	S12

T14.0	S15	*T28.7*	D80	*T70.0*	H79	*T88.1*	A85
T14.0	S16	*T28.8*	U80	*T70.1*	R88	*T88.2*	A85
T14.0	S17	*T28.8*	X82	*T70.2*	A88	*T88.3*	A85
T14.0	S19	*T28.8*	Y80	*T70.3*	A88	*T88.4*	A85
T14.1	S13	*T28.9*	A80	*T70.4*	A88	*T88.5*	A85
T14.1	S15	*T29*	A81	*T70.8*	A88	*T88.6*	A87
T14.1	S18	*T30*	S14	*T70.9*	A88	*T88.7*	A87
T14.2	L76	*T31*	S14	*T71*	A88	*T88.8*	A85
T14.3	L79	*T32*	S14	*T73*	A88	*T88.9*	A85
T14.3	L80	*T33*	A88	*T74.0*	Z12	*T90*	A82
T14.4	N81	*T34*	A88	*T74.0*	Z16	*T91*	A82
T14.5	A80	*T35*	A88	*T74.0*	Z20	*T92*	A82
T14.6	L81	*T36*	A84	*T74.1*	Z25	*T93*	A82
T14.7	A80	*T37*	A84	*T74.2*	Z25	*T94*	A82
T14.7	L81	*T38*	A84	*T74.3*	Z12	*T95*	A82
T14.8	A80	*T39*	A84	*T74.3*	Z16	*T96*	A82
T14.9	A80	*T40*	A84	*T74.8*	Z25	*T97*	A82
T14.9	B77	*T41*	A84	*T74.9*	Z25	*T98*	A82
T14.9	W75	*T42*	A84	*T75*	A88	*Z00*	–30
T15	F76	*T43*	A84	*T78.0*	A92	*Z00*	A97
T16	H76	*T44*	A84	*T78.1*	A92	*Z01*	–31
T17	R87	*T45*	A84	*T78.2*	A92	*Z01*	–32
T18	D79	*T46*	A84	*T78.3*	A92	*Z01*	–33
T19.0	U80	*T47*	A84	*T78.4*	A92	*Z01*	–34
T19.1	U80	*T48*	A84	*T78.8*	A88	*Z01*	–35
T19.2	X82	*T49*	A84	*T78.9*	A88	*Z01*	–36
T19.3	X82	*T50*	A84	*T79.0*	A82	*Z01*	–37
T19.8	U99	*T51*	A86	*T79.1*	A82	*Z01*	–38
T19.9	U99	*T52*	A86	*T79.2*	A82	*Z01*	–39
T20	S14	*T53*	A86	*T79.3*	S11	*Z01*	–40
T21	S14	*T54*	A86	*T79.4*	A82	*Z01*	–41
T22	S14	*T55*	A86	*T79.5*	A82	*Z01*	–42
T23	S14	*T56*	A86	*T79.6*	L99	*Z01*	–43
T24	S14	*T57*	A86	*T79.7*	A82	*Z02*	–30
T25	S14	*T58*	A86	*T79.8*	A82	*Z02*	–62
T26	F79	*T59*	A86	*T79.9*	A82	*Z03*	–45
T27	R88	*T60*	A86	*T80*	A87	*Z04.0*	–34
T28.0	D80	*T61*	A86	*T81*	A87	*Z04.1*	–31
T28.1	D80	*T62*	A86	*T82*	A89	*Z04.2*	–31
T28.2	D80	*T63*	A86	*T83*	A89	*Z04.3*	–31
T28.3	U80	*T64*	A86	*T83*	W12	*Z04.4*	–31
T28.3	X82	*T65*	A86	*T84*	A89	*Z04.5*	–31
T28.3	Y80	*T66*	A88	*T85*	A89	*Z04.6*	–31
T28.4	A80	*T67*	A88	*T86*	A87	*Z04.8*	–31
T28.5	D80	*T68*	A88	*T87*	A87	*Z04.9*	–31
T28.6	D80	*T69*	A88	*T88.0*	A85	*Z08*	–31

Z09	–31	*Z30.3*	W10	*Z46.3*	D54	*Z51.8*	–69
Z10	–30	*Z30.3*	W83	*Z46.4*	D54	*Z51.8*	A98
Z11	–31	*Z30.4*	W11	*Z46.5*	D54	*Z51.9*	–59
Z11	–32	*Z30.5*	W12	*Z46.6*	U54	*Z51.9*	–63
Z11	–33	*Z30.8*	–37	*Z46.7*	L54	*Z51.9*	–64
Z11	–34	*Z30.8*	W14	*Z46.8*	–54	*Z51.9*	–65
Z11	–35	*Z30.8*	Y14	*Z46.9*	A54	*Z51.9*	–66
Z11	–36	*Z30.9*	W14	*Z47*	L54	*Z51.9*	–67
Z11	–38	*Z30.9*	X10	*Z48*	–51	*Z51.9*	–68
Z11	–39	*Z30.9*	Y14	*Z48*	–52	*Z51.9*	–69
Z12	–31	*Z31*	W15	*Z48*	–54	*Z52*	–69
Z12	–32	*Z32.0*	W01	*Z48*	–56	*Z53*	–69
Z12	–33	*Z32.1*	W78	*Z48*	–59	*Z54*	–63
Z12	–34	*Z32.1*	W79	*Z49*	U51	*Z54*	–69
Z12	–35	*Z33*	W78	*Z49*	U59	*Z55*	Z07
Z12	–36	*Z34*	W31	*Z50*	–57	*Z56.0*	Z06
Z12	–37	*Z34*	W78	*Z51.0*	–59	*Z56.1*	Z05
Z12	–38	*Z35*	W84	*Z51.1*	–50	*Z56.2*	Z05
Z12	–40	*Z36*	–43	*Z51.2*	–50	*Z56.3*	Z05
Z12	–41	*Z37.0*	W90	*Z51.3*	–50	*Z56.4*	Z05
Z12	–43	*Z37.1*	W91	*Z51.4*	–59	*Z56.5*	Z05
Z13	–31	*Z37.1*	W93	*Z51.5*	–50	*Z56.6*	Z05
Z13	–34	*Z37.2*	W92	*Z51.5*	–51	*Z56.7*	Z05
Z13	–35	*Z37.3*	W93	*Z51.5*	–53	*Z57*	Z05
Z13	–36	*Z37.4*	W93	*Z51.5*	–55	*Z58.0*	Z29
Z13	–38	*Z37.5*	W92	*Z51.5*	–57	*Z58.1*	Z29
Z13	–39	*Z37.6*	W93	*Z51.5*	–58	*Z58.2*	Z29
Z13	–40	*Z37.7*	W93	*Z51.5*	–59	*Z58.3*	Z29
Z13	–41	*Z37.9*	W90	*Z51.6*	–55	*Z58.4*	Z29
Z13	–42	*Z37.9*	W91	*Z51.8*	–50	*Z58.5*	Z29
Z13	–43	*Z37.9*	W92	*Z51.8*	–50	*Z58.6*	Z02
Z20	A23	*Z37.9*	W93	*Z51.8*	–51	*Z58.8*	Z29
Z21	B90	*Z38*	W90	*Z51.8*	–52	*Z58.9*	Z29
Z22	A99	*Z38*	W92	*Z51.8*	–53	*Z59.0*	Z03
Z23	A23	*Z39*	W31	*Z51.8*	–54	*Z59.1*	Z03
Z24	A23	*Z40*	–59	*Z51.8*	–55	*Z59.2*	Z03
Z25	A23	*Z41*	–69	*Z51.8*	–56	*Z59.3*	Z03
Z26	A23	*Z42*	–59	*Z51.8*	–57	*Z59.4*	Z02
Z27	A23	*Z43*	–53	*Z51.8*	–58	*Z59.5*	Z01
Z28	A23	*Z43*	–54	*Z51.8*	–62	*Z59.6*	Z01
Z29	A23	*Z44*	–54	*Z51.8*	–63	*Z59.7*	Z01
Z30.0	W14	*Z45*	–54	*Z51.8*	–64	*Z59.7*	Z08
Z30.0	Y14	*Z46.0*	F17	*Z51.8*	–65	*Z59.8*	Z01
Z30.1	W12	*Z46.0*	F18	*Z51.8*	–66	*Z59.8*	Z03
Z30.2	W13	*Z46.1*	H54	*Z51.8*	–67	*Z59.9*	Z01
Z30.2	Y13	*Z46.2*	N54	*Z51.8*	–68	*Z59.9*	Z03

Z60	P25
Z60	Z04
Z60.0	P25
Z61	Z16
Z62	Z16
Z63.0	Z12
Z63.0	Z13
Z63.1	Z20
Z63.1	Z21
Z63.2	Z29
Z63.3	Z29
Z63.4	Z15
Z63.4	Z19
Z63.4	Z23
Z63.5	Z15
Z63.6	Z14
Z63.6	Z18
Z63.6	Z22
Z63.7	Z22
Z63.7	Z29
Z63.8	Z16
Z63.8	Z20
Z63.8	Z29
Z63.9	Z21
Z63.9	Z24
Z63.9	Z29
Z64.0	Z29
Z64.1	Z29
Z64.2	P29
Z64.3	P29
Z64.4	Z10
Z65.0	Z09
Z65.1	Z09
Z65.2	Z09
Z65.3	Z09
Z65.4	Z25
Z65.5	Z25
Z65.8	Z29
Z65.9	Z29
Z70	P58
Z70	X24
Z71.0	–65
Z71.0	–69
Z71.1	–69
Z71.1	A13
Z71.1	A25
Z71.1	A26
Z71.1	A27
Z71.1	B25
Z71.1	B26
Z71.1	B27
Z71.1	D26
Z71.1	D27
Z71.1	F27
Z71.1	H27
Z71.1	K24
Z71.1	K25
Z71.1	K27
Z71.1	L26
Z71.1	L27
Z71.1	N26
Z71.1	N27
Z71.1	P27
Z71.1	R26
Z71.1	R27
Z71.1	S26
Z71.1	S27
Z71.1	T26
Z71.1	T27
Z71.1	U26
Z71.1	U27
Z71.1	W02
Z71.1	W27
Z71.1	X23
Z71.1	X24
Z71.1	X25
Z71.1	X26
Z71.1	X27
Z71.1	Y24
Z71.1	Y25
Z71.1	Y26
Z71.1	Y27
Z71.1	Z27
Z71.2	–60
Z71.2	–61
Z71.3	–45
Z71.4	P58
Z71.5	P58
Z71.6	P58
Z71.7	B58
Z71.8	–58
Z71.8	A20
Z71.9	–45
Z71.9	–58
Z72.0	A23
Z72.1	A23
Z72.2	A23
Z72.3	A23
Z72.4	A23
Z72.5	A23
Z72.6	Z29
Z72.8	Z29
Z72.9	Z29
Z73.0	P29
Z73.1	P29
Z73.2	A23
Z73.2	Z29
Z73.4	Z28
Z73.5	Z29
Z73.6	–28
Z73.6	A28
Z73.6	B28
Z73.6	D28
Z73.6	F28
Z73.6	H28
Z73.6	K28
Z73.6	L28
Z73.6	N28
Z73.6	P28
Z73.6	R28
Z73.6	S28
Z73.6	T28
Z73.6	U28
Z73.6	W28
Z73.6	X28
Z73.6	Y28
Z73.6	Z28
Z73.8	Z29
Z73.9	Z29
Z74	A28
Z75	Z10
Z75	Z11
Z76.0	–50
Z76.1	Z29
Z76.2	–49
Z76.3	–69
Z76.4	–69
Z76.5	P80
Z76.8	–46
Z76.8	–47
Z76.8	–48
Z76.8	–49
Z76.8	–64
Z76.8	–65
Z76.8	–69
Z76.8	A97
Z76.9	–69
Z80	A21
Z81	A23
Z82.0	A23
Z82.1	A23
Z82.2	A23
Z82.3	K22
Z82.4	K22
Z82.5	A23
Z82.6	A23
Z82.7	A23
Z82.8	A23
Z83	A23
Z84	A23
Z85	A21
Z86.0	A23
Z86.1	A23
Z86.2	A23
Z86.3	A23
Z86.4	A23
Z86.5	A23
Z86.6	A23
Z86.7	K22
Z87	A23
Z88	A23
Z89	L99
Z90.0	A99
Z90.1	X99
Z90.2	R99
Z90.3	D99
Z90.4	D99
Z90.5	U99
Z90.6	U99
Z90.7	X28
Z90.7	X99
Z90.7	Y28
Z90.7	Y99
Z90.8	A99

Z91	A23	*Z93.8*	A28	*Z94.6*	L99	*Z98.1*	L99
Z92	A23	*Z93.9*	A28	*Z94.7*	F99	*Z98.2*	N99
Z93.0	R28	*Z94.0*	U99	*Z94.8*	A99	*Z98.8*	A99
Z93.1	D28	*Z94.1*	K99	*Z94.9*	–99	*Z99.0*	A28
Z93.2	D28	*Z94.2*	R99	*Z94.9*	A99	*Z99.1*	R28
Z93.3	D28	*Z94.3*	K99	*Z95*	A89	*Z99.2*	U28
Z93.4	D28	*Z94.3*	R99	*Z96*	A89	*Z99.3*	A28
Z93.5	U28	*Z94.4*	D99	*Z97*	A89	*Z99.8*	A28
Z93.6	U28	*Z94.5*	S99	*Z98.0*	D99	*Z99.9*	A28

12 Alphabetical index

This index is not meant to be comprehensive, nor to be a nomenclature (see Chapter 1). It is a list only of the titles of rubrics (in upper case) and of inclusion terms in the rubrics (in lower case). These comprise the synonyms and terms most commonly used in general/ family practice. Users requiring a more extensive index or nomenclature will need to develop their own or use ones already available in countries such as Australia, Canada, The Netherlands, and some Scandinavian countries. In order to maintain consistency, this should be done in cooperation with the WONCA Classification Committee.

Abbreviations are not included in this index, except the following:
complt = complaint
/ = or
sympt = symptom
NOS = not otherwise specified in this classification